BEST OF
PHYSICS
FROM

Science Teacher's Workshop

BEST OF
PHYSICS
FROM

Compiled and Edited by
The Board of Editors of the

Parker Publishing Co., Inc.

Science Teacher's Workshop

Science Teacher's Workshop

West Nyack, New York

Best of Physics from
Science Teacher's Workshop

Compiled and Edited by

The Board of Editors of the
Science Teacher's Workshop

© 1972, by

PARKER PUBLISHING COMPANY, INC.
West Nyack, New York

Library of Congress
Catalog Card Number: 70-163639

Printed in the United States of America
ISBN 0-13-073874-3
B & P

ABOUT THIS BOOK

This book offers you, the physics teacher, a store of innovative, practical demonstrations for teaching high school physics. Each one is complete and ready for use in teaching some aspect of the modern course, from Kinematics and Newton's Laws of Motion, Light and Wave Behavior, to Electricity and Electronics. Together, they represent a unique collection of ideas and techniques for demonstrating traditional subject matter as well as recent developments on the frontiers of physics.

The materials have been specially selected from articles written by experienced physics teachers for the *Science Teacher's Workshop,* a monthly educational service for secondary-school science teachers. All of them have been tested with students and found particularly effective in stimulating students' interest and in helping them to understand better the laws of physics and their workings in our world.

Practicableness as well as effectiveness has figured in the selection of the demonstrations. Very few require unusual equipment or involve elaborate procedures. Rather, they represent the efforts of many physics teachers to find the most useful and vivid means within their situations, to illustrate the concepts and processes of physics to their students.

As an aid to you, an effort has been made to furnish all of the details you will need to carry out each demonstration activity successfully in your own classroom or laboratory. For easy and effective use of ideas:

  *Objectives are spelled out and materials are conveniently listed.
  *Sources or substitutes are provided for equipment not usually found in the high school physics lab.
  *Procedures for the teacher and his students are clearly presented in step-by-step fashion.
  *Special notes are included to help avoid difficulties and to stress safe practices.
  *Sample analyses and calculations are included to serve as a useful guideline to development.

5

*Follow-up activities are suggested to reinforce learning and spur further student investigation.

*Illustrations are used to provide a concrete reference to procedures.

This book should help you to vary, enrich, and perhaps even replace some parts of your regular physics teaching program. It will save you time, energy, and expense, and many of the techniques presented will help you overcome difficulties in teaching physics that have consistently been a problem in the past.

We hope this book will make the study of physics more rewarding and more fun for both you and your students.

Board of Editors
Science Teacher's Workshop

CONTENTS

8. PERIODIC MOTION AND SPACE PHYSICS

9. CONSERVATION OF MOMENTUM

10. WORK AND ENERGY

11. STATIC ELECTRICITY

BEST OF PHYSICS FROM

Science Teacher's Workshop

1

MATHEMATICS AND MEASUREMENT

MEASUREMENTS IN THE PHYSICS LABORATORY

by David Kutliroff [*]

New Brunswick Senior High School, New Brunswick, N.J.

It is important to emphasize at the start of the physics course that it is a quantitative science. The physics laboratory is a place to make measurements. The necessity to base much of our understanding of the physical sciences on indirect measurements justifies spending some time at the beginning of an introductory course in physics, familiarizing students with some of the techniques.

Here are several simple laboratory exercises in different areas of measurements:

Triangulations

Take students out to the school parking lot or athletic field and ask them to measure the distance between two distant objects without actually laying out a tape. You may select two trees, a car and a telephone pole, and so on.

NOTE: Going outside for an early lab emphasizes the idea that a laboratory is not necessarily a room filled with stainless steel instruments. It can be any place where measurements are made.

A base line will be selected and measurements made by triangulation. Materials needed:

Cork board (or soft wood board which can hold straight pins)
Paper and pencil
Straight pins
Meter stick

NOTE: Some laboratory exercises use simple range finders, parallax viewers, and so on. At an early stage, it is well to emphasize that these are just *methods* of triangulation; the student should understand what he is doing rather than rely on working with what to him, at that point, is a "black box" instrument.

---

*Mr. Kutliroff is the author of a new book for secondary-school physics teachers, *Physics Teacher's Guide: Effective Classroom Demonstrations and Activities* (West Nyack, N. Y.: Parker Publishing Company, Inc., 1970).

Two students can work on this problem. Their cork board with a paper tacked on to it must always have one side parallel to the base line and a corresponding base line drawn on the paper parallel to that side. One student then stands on a spot on the base line which he marks and calls "A." He marks a similar point on the base line on the paper and marks it "A^1." He then sights with straight pins to the two distant objects, thus constructing the viewing angles at one end of his base line between the base line and the lines of sight to the two distant objects. (See Fig. 1.)

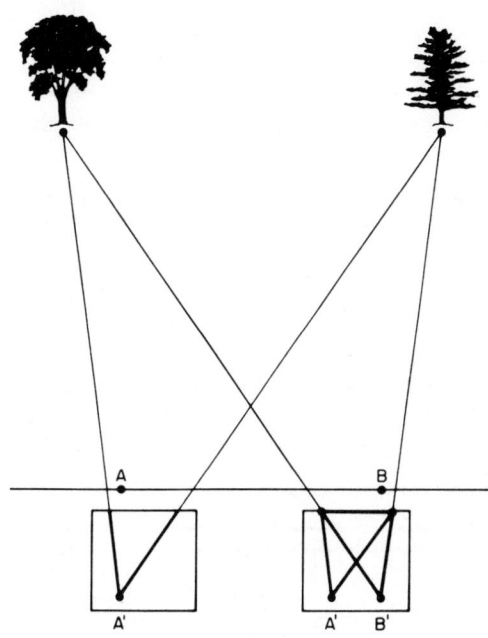

Fig. 1

The partner's function at this point is to make sure that the base line on the sighting tablet is always parallel to the actual base line while the first student is constructing the angles. They then walk over to the other end of their selected base line, which can be called "B," and mark point "B^1" on the corresponding base line on the sighting pad. The sight lines are drawn and the sighting angles are constructed similarly to the way the sight lines were found from point A.

The sight lines from both ends of the base line are then extended until they intersect. The intersection of the sight lines on the paper gives the positions of the two objects sighted in real space.

Since simple geometry can indicate that all the figures on the paper are similar to the geometry of the actual spaces measured, and corresponding parts of similar figures have the same ratio, the student can, by measuring base lines A^1–B^1 on the paper and the actual base line A–B from whose ends he made the sightings, determine the distance between the two distant objects to which he sighted, or the distance from either of these objects to any point on the base line.

NOTE: After completing his figure, the student ought to be encouraged to make the direct measurement to check himself and calculate the per cent of error.

Irregular Areas

You can ask your students, for example, to find the area of the county in which they live and to use two different methods of computing the area.

NOTE: This could be assigned as a homework lab. This is a good practice once in a while as it again emphasizes that the location of a laboratory is unimportant so long as it is a convenient place to make measurements

Method 1: Obtain a road map and cut out the county whose area is being determined. Trace the outline of the county on a piece of graph paper, and determine the area represented by each square by the scale given on the map. The area of the county can then be determined by counting squares and estimating partial sequences.

Method 2: You can also determine the area by *weighing it.* The first step is the construction of a simple, sensitive equal arm balance with a soda straw, straight pin, razor blades, and a little aluminum foil. (See Fig. 2.)

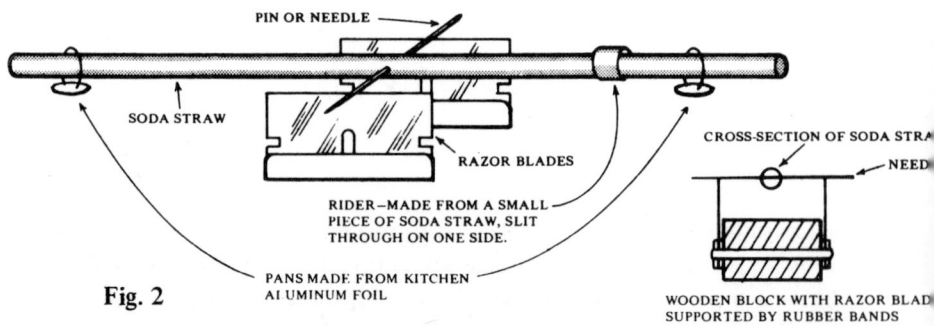

Fig. 2

NOTE: The balance once recommended in the very early versions of PSSC laboratory manual would be quite satisfactory.

In this method, we assume, of course, that the road map is made of a fairly homogeneous paper of common thickness and density. The cut-out section of the map is placed on a pan of the balance, and cut-out squares of graph paper are then placed on the other pan until they balance. Then a square of road map paper, whose area is known, because of the marked scale on the map, is placed on the scale and balanced similarly with cut-out squares of graph paper. The area can be determined because the ratio of their areas is equal to the ratio of their weights.

Exponential Notation

In physics, the students will handle very large numbers; they may, for example, talk about distances to stars wherein light takes a million or more years to get to us. Thus, if light travels 300,000,000 meters per second and there are 3600 seconds in an hour, 24 hours in a day, and 365 days in a year, going through calculations to find the number of meters in one light year and just writing down the final number would be laborious indeed. Thus, the students should know how to use exponential notations.

Instead of writing 300,000,000, we can write 3.0×10^8. This means 3 x (10 multiplied by itself 8 times). Similarly, 3600 can be written 3.6×10^3. Thus, instead of multiplying 300,000,000 by 3600, we can:

Multiply 3×10^8 by 3.6×10^3, or
$3.0 \times 3.6 \times 10^8 \times 10^3$
$3.0 \times 3.6 = 10.8$ and $10^8 \times 10^3 = 10^{11}$.

NOTE: In multiplying powers of ten, we merely add the exponents. Thus, ten multiplied by itself eight times, times ten multiplied by itself three times, equals ten multiplied by itself 11 times.

Therefore, $(3.0 \times 10^8) \times (3.6 \times 10^3) = 10.8 \times 10^{11}$

Let us go through the entire calculation:

3.0×10^8 (300,000,000 meters) x 3.6×10^3 (3600 seconds) x 2.4×10^1 (24 hours) x 3.7×10^2 (365 days, leveled out to 370). It works out this way:

$3.0 \times 10^8 \times 3.6 \times 10^3 \times 2.4 \times 10^1 \times 3.7 \times 10^2 = 95.90 \times 10^{14} = 96 \times 10^{14} = 9.6 \times 10^{15}$ meters.

Think of handling 96 with 14 zeros after it!

At the other end, students will have to consider very small numbers. For example, the average distance between molecules in a room is $1/10^8$ meters. Let us put down a few equivalents so you can see how the notation works:

$$\frac{1}{10} = \frac{1}{10^1} = .1 = 10^{-1}$$

$$\frac{1}{100} = \frac{1}{10^2} = .01 = 10^{-2}$$

$$\frac{1}{1000} = \frac{1}{10^3} = .001 = 10^{-3}.$$

If we multiply $\frac{1 \times 1000}{10}$, we get 100. Similarly, therefore: $10^{-1} \times 10^3 = 10^2$. We are still multiplying by algebraically adding exponents, because $-1 + 3 = 2$.

DEMONSTRATING THE ROLE
OF MATHEMATICS IN SCIENCE

by D.W. Tomer

Hanford Joint Union High School, Hanford, California

To the scientist, mathematics is an analytic tool applied to experimental data with the hope of cranking out a formula that describes, at least approximately, some basic tendency of nature. I have devised an exercise that illuminates the role of math in discovering and describing the physical world.

NOTE: The method described here has been successfully used at the eleventh and twelfth grade levels, and may be successfully used in classes as low as the seventh grade.

Materials and Preparation

For the particular exercise described here, you will need:

1 meter of heavy, bare, solid metal wire
Wire cutters
Weighing scale
Centimeter scales for each student
Graph paper for each student.

Before class, carefully balance the scales. Cut a piece of wire and fasten it to the bottom of the weighing pan with a minimum of Scotch tape. This piece of wire must be concealed from the class; it represents an unsuspected fact in nature.

NOTE: Some may think that this deception is not sound at this level. I believe it illustrates how unknown and unsuspected facts upset experimental results, and the deception has an essential role in science education.

When the class is assembled, pass out the centimeter scales and graph paper, explaining that this is an exercise illustrating the problems of scientists in applying mathematics to research. Show the length of wire, whose properties are to be investigated, then cut off at random five or six pieces of widely differing lengths. Hand one piece of wire to the first student in each row, asking him to measure it carefully and pass it to the second student, and so on, until all students in the row have measured it and come to an agreement as to the most probable value of its length. As soon as the students are in agreement, collect the pieces and weigh each *except the last*. Tabulate the results on the board.

This might result in the following set of data:

Wire	1	2	3	4	5
Centimeters	4.00	6.50	5.48	1.51	3.12
Grams	1.23	1.72	1.51	.79	

The Problem

The problem is now obvious to the students: Predict the missing weight. You may wish to let class members suggest ways of doing this, but it may ultimately be necessary for you to suggest graphing. For some classes, it is necessary to spend a previous class period on graphing so that the problem at hand does not become lost in the technicalities of making a graph.

If everything has been carefully done, the plotted points will fall in a nearly straight line (see graph, Fig. 1). These questions may come up:

(1) Some students will want to plot straight-line segments between each pair of adjacent points. Explain that broken lines are very difficult to handle mathematically, and just as scientists often start by simplifying a complicated problem, the student is wise to plot a straight line of best fit. (The line should be extended to the left across both axes, as shown in Fig. 1.)

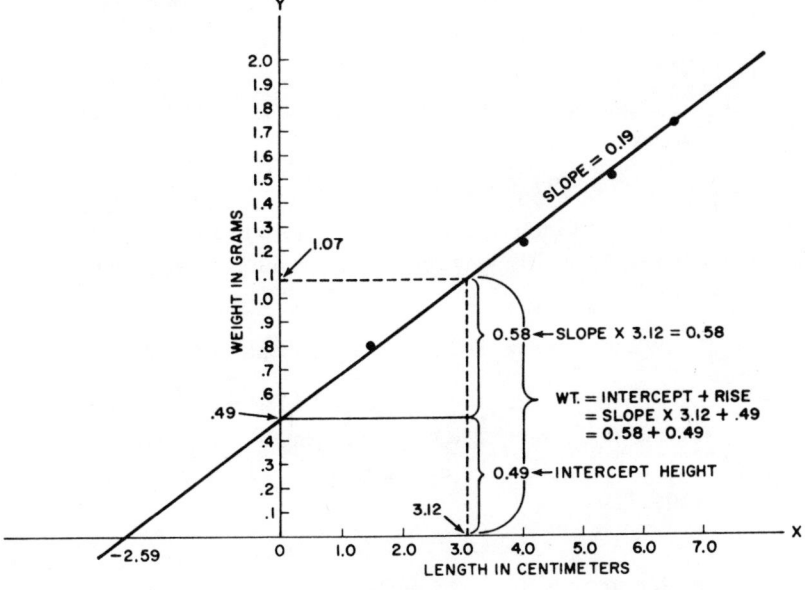

Fig. 1

(2) Occasionally a student, reasoning that a wire of zero length would have a zero weight, will want to curve the line down through the origin. Congratulate the good thinking, but explain that it contradicts the equally plausible assumption that uniform wire should exhibit a uniform change in weight for each change in length (constant slope → straight line), and since zero length and weight are not part of the data, it is better to stick to the assumption that keeps the problem simple. Tell the student to go along with the straight line and the difficulty will soon be resolved.

When the graphs are complete, most students will be able to predict quickly the required weight by interpolation. Let them compare results and come to an agreement on the most probable value. Immediately thereafter, weigh the wire (on the same loaded balance, of course). While *exact* agreement between the predicted and measured values will be attained only by accident, the agreement will usually be close enough to convince even the most skeptical student. Faith in the graph will then permit you to derive from it a more general result; namely, a formula.

Deriving the Formula

Point out that, on the graph, a vertical distance represents a weight whose magnitude can be determined by comparison with the scale along the Y axis. Thus, in the example under consideration, the students located the length of the unknown (3.12) on the X axis and found the vertical distance up to the straight line was equivalent to 1.07 on the weight scale on the Y axis. Therefore, 1.07 grams may be represented as the sum of two parts:

(1) The height of the intercept on the Y axis, or 0.49 grams in this case.
(2) The amount the line rises in going from zero to 3.12 cm, or 0.58 grams.

In general terms: Weight=Intercept + Rise.

The value of the intercept may be read directly from the graph. The rise may be computed by multiplying the length by the rise per unit run (slope) of the line. Thus:

$$\text{Weight} = b + m \times \text{Length} \quad (b = \text{intercept}; m = \text{slope}).$$

To find the slope, the simplest method is to pick two extreme points, construct and measure $\triangle X$ (weight) and $\triangle Y$ (length), and divide. In the illustration used, you will find that the line rises close to 0.19 grams per unit on the centimeter scale. Thus:

$$\text{Slope} = 0.19 \text{ grams/cm.}$$

The specific formula for the example used here then becomes:

$$\text{Grams} = 0.49 + 0.19 \times \text{centimeters.}$$

At this point, let the students practice with the formula. Cut a different length of wire, measure it, and let the students predict its weight by both graph and formula, and check it on the scales. Emphasize that the formula is found from the graph and says the same thing as the graph. Argue that the formula is more useful than the graph, since it is more compact, easier to remember, and extends beyond the range of the graph.

Finally, focus attention on the constants, 0.49 and 0.19. Since these are derived from physical measurements, they may have physical meaning; i.e., they may describe something real. What do they describe? For a clue, look at their labels.

The 0.19 was obtained by dividing weight by length ($\triangle W/\triangle L$), so its proper label is grams/centimeters. It is the linear density of the wire, or the amount the weight increases for each additional centimeter of length.

The 0.49 appears as a point on the Y axis or weight scale. It must be the weight of something. The line also crosses the X axis at -2.59 centimeters. Could this be the length of something? Where is there something that is 2.59 centimeters long and weighs 0.49 grams?

It is well to provide a little time for the students to form hypotheses. Do not evaluate any of these; just encourage them to propose as many as possible. When the ideas run out, reveal that although you balanced the scales initially, you next unbalanced them with a small piece of wire that has a length of—guess how much? (The weight of the Scotch tape usually does not matter.)

Summarizing, the mathematical analysis produced a description of one of the wire's properties. It also found the length and weight of an unsuspected piece of wire.

NOTE: Identifying the last relation is quite difficult; it is a rare student who will suspect this form of deception. Point out, however, that nature is often deceptive, and when the experimentalist, Boyle, discovered the gas law PV = K, it was a rare theoretician, Bernoulli, who suspected the nature of its cause.

RELATIONS BETWEEN
KINEMATICAL GRAPHS

by George F. Smith

South Hadley High School, South Hadley, Massachusetts

Most beginning physics students have difficulty understanding the relationships between functions involving distance time, speed time, and acceleration time. These ideas are new to them and they have not had adequate background experiences working with these concepts and their interrelations.

This demonstration-experiment is designed to provide students with some of the needed experience. If several similar experiences are provided for beginning students, they will create a firm base for further study of kinematics.

NOTE: Prior to attempting the experiment, students should have had some experience involving the use of apparatus for recording distance vs. time—for instance, PSSC type carts and recording timers. They should also be familiar with the construction of a distance-time graph from recorded data.

Experimental Procedure

The procedure is as follows:

(1) Incline a table at approximately 30 degrees with the horizontal by putting a support under one end, and allow a cart to roll down the incline (Fig. 1).

(2) Obtain a distance-time record of the trip by attaching some PSSC type of recording tape to the cart and allowing the cart to pass through some PSSC type of recording timer. (The resulting series of

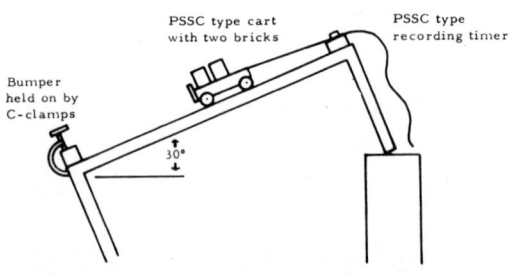

Fig. 1

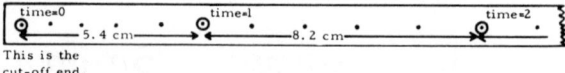

Fig. 2

dots on the tape yields a permanent record of the motion of the cart down the inclined surface.)

(3)   Cut off the first few centimeters of the tape so that the remaining trip record will be for the cart already in motion.

(4)   Circle the first dot on the tape and then every fifth dot (Fig. 2); label the circled dots as time = 0, time = 1, etc.

(5)   Measure the distance between the circled dots and record the distance directly on the tape.

(6)   From this tape, develop a table of time, distance, speed, and acceleration (Table 1).

Table 1
Original Data (Graphs 1, 2, and 3)

Time (Intervals of 5 dots)	Total distance accomplished (cm)	Average speed during interval (cm/ 5 dots)	Acceleration (cm/(5 dots)^2)
0	0		
		5.4	
1	5.4		2.8
		8.2	
2	13.6		2.6
		10.8	
3	24.4		2.9
		13.7	
4	38.1		2.7
		16.4	
5	54.5		3.1
		19.5	
6	75.0		3.2
		21.7	
7	96.7		3.1
		24.8	
8	121.5		

Average=2.9 cm/(5 dots)^2

The length of each interval of five dots represents the average speed during the interval (the time for five dots will be the standard unit of time throughout the experiment); the total distance accomplished is a running total of the lengths of these intervals, and the acceleration is the change in speed that takes place from one interval to the next.

NOTE: This table of time, distance, speed, and acceleration provides all the data necessary to construct a distance-time graph (Fig. 3, Graph 1), a speed-time graph (Fig. 4, Graph 2), and acceleration-time graph (Fig. 5, Graph 3). Since the speeds represent averages between designated times, they are plotted midway between the appropriate times.

Reconstructing Graphs

The remainder of the experiment is concerned with the problems of reconstructing these graphs when knowing only one of them. The speed at any given time can be determined from the distance-time graph (Fig. 3, Graph 1) by constructing a tangent to the curve at the given time. The slope of this tangent line will be the same as the speed at that time.

EXAMPLE: The slope of Graph 1 at time = 4 is 15 cm/ (5 dots), which is the same as the speed given on the speed-time graph (Fig. 4, Graph 2) for the same time.

Students should try this slope process several times to see how easy it is to derive speeds from the distance-time graph. In a similar fashion, they should find the acceleration at a few different times, using the slope of the speed-time graph (Graph 2), and verify their results by checking the acceleration-time graph (Fig. 5, Graph 3).

It is a little more difficult to go from the acceleration-time graph to the speed-time graph and from the speed-time graph to the distance-time graph. The general rule is that the cumulative area under the graph represents the net change in some quantity since the beginning of the problem.

EXAMPLE: Using the acceleration-time graph, the area from time = 0 to time = 2 is 2 (2.9) = 5.8 cm/ (5 dots). If the original speed is assumed to be zero, then the total speed at time = 2 is 0 + 5.8 = 5.8 cm/ (5 dots).

In a similar fashion, the students can construct an entire speed-time graph (Fig. 4, Graph 4). The teacher should point out to students that this graph (Graph 4) is similar to the experimental one (Graph 2) but is not identical. In fact, it seems to be the same as the experimental one except that it is displaced in the vertical direction by about 4 cm/ (5 dots).

NOTE: The area technique produced graphs which are displaced in the vertical direction. In order to obtain the experimental graph, it is

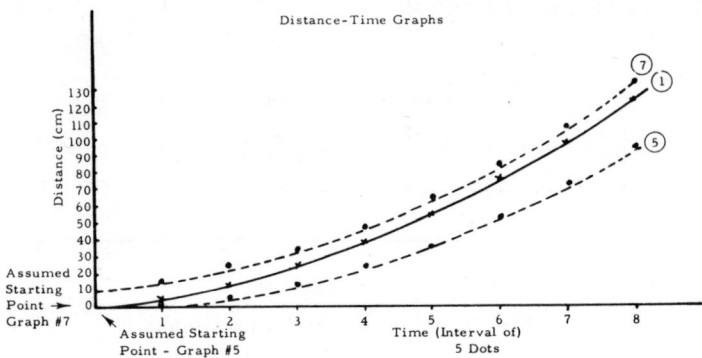

Fig. 3: Distance-Time Graphs

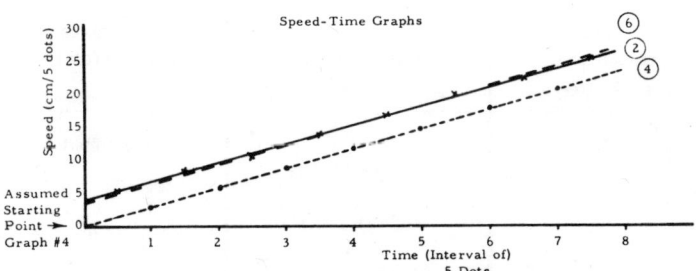

Fig. 4: Speed-Time Graphs

Table 2
Computations for Graph 5 (Fig. 3)

Time (Intervals of 5 dots)	Distance covered during interval (cm)	Total distance covered (cm)
0		0 assumed
	(0+3)/2=1.5	
1		1.5
	(3+6)/2=4.5	
2		6.0
	(6+8.5)/2=7.3	
3		13
	(8.5+11.5)/2=10	
4		23
	(11.5+14.5)/2=13	
5		36
	(14.5+17.5)/2=16	
6		52
	(17.5+20.5)/2=19	
7		71
	(20.5+23.5)/2=22	
8		93

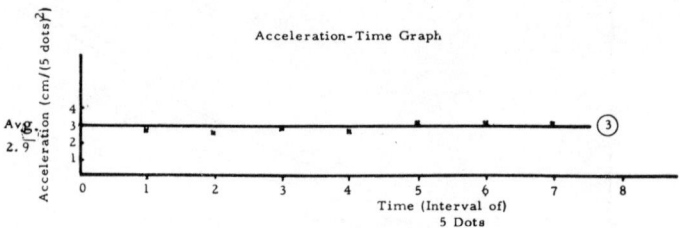

Fig. 5: Acceleration-Time Graphs

Table 3
Computations for Graph 7 (Fig. 3)

Time (Intervals of 5 dots)	Distance covered during interval (cm)	Total distance covered (cm)
0		10 assumed
	(3.5+6.5)/2=5	
1		15
	(6.5+9)/2=7.8	
2		23
	(9+12)/2=10.5	
3		33
	(12+15)/2=13.5	
4		47
	(15+18)/2=16.5	
5		63
	(18+21)/2=19.5	
6		83
	(21+24)/2=22.5	
7		105
	(24+27)/2=25.5	
8		131

necessary to know the correct experimental speed for some time so that the vertical displacement correction can be applied to the derived graph (Fig. 4, Graph 4).

Students should try making a table (Table 2) of cumulative areas for various times, using the speed-time graph (Graph 4) just derived. When the derived cumulative distance data are plotted, a distance-time graph (Fig. 3, Graph 5) can be constructed. As before, the areas indicate only the cumulative change since the beginning of the problem, so an original total distance must be assumed (in Graph 5, it was assumed that the total distance = 0 when time = 0). Obviously, Graph 5 does not look like the experimental graph (Graph 1). It is not even the same graph slightly displaced in a vertical direction.

The teacher should encourage students to speculate as to what went wrong. Sooner or later someone will suggest that Graph 4 was not really correct, so that anything derived from it will probably be in error. This opens the door to displacing Graph 4 vertically so that it matches an experimental speed at some given time (Graph 6).

Using this new, corrected speed-time graph (Graph 6), the area principle should be tried again, assuming some given distance at time = 0 (Table 3 assumes that the total distance = 10 cm when time = 0). Once again, when the data is plotted (Fig. 3, Graph 7), a graph is obtained which is vertically displaced from the experimental one (Graph 1).

Using the Exercises

The types of exercises presented here can be used to convey a number of ideas to the students. When using the slope method to derive a succeeding graph, good results are obtained and no information needs to be given other than the original graph. However, when using the area method, the area represents only the net change since the beginning of the problem. The result of plotting net changes since the beginning of the problem is a graph which is displaced vertically from the experimental graph. In order to correct the derived graph, at least one point of the experimental graph must be known so that a vertical adjustment can be made. The area method cannot be used successively in constructing graphs unless vertical corrections are made at each step.

> **NOTE:** Teachers will recognize that the methods used in this experiment are equivalent to those used for differentiation and integration, but do not require a calculus background. They will also recognize that the problem of the necessary vertical displacement corrections is essentially the same as that of constants of integration.

I believe that the basic ideas of distance, speed, and acceleration, along with their interrelations, are very difficult for the average high school physics student to comprehend, and that teachers should give careful consideration to the development of a proper experience foundation before they expect students to become really operational with these concepts. Experiments such as this one can contribute to a firm background for the study of kinetics.

A LABORATORY APPROACH
TO RECTILINEAR KINEMATICS

by Lawrence J. Badar

Rocky River High School, Rocky River, Ohio

In order to better prepare students for the new science courses (BSCS, PSSC, CHEM Study), we cover relatively few topics in our ninth grade science courses, but go over them in depth. The course is grounded in the laboratory, and we follow a rule of "guided discovery"; i.e., experimental exercises presented as problems to be solved rather than recipes to be followed.

One example is the handling of rectilinear kinematics. Every aspect of this simple motion is subject to laboratory discovery, extensive graphing, and mathematical analysis of the results. Here are the important points.

1. We start with a very common motion—walking. Students mark off a section of sidewalk (or corridor) in 1-meter lengths. Then each pair of students is assigned a time interval (usually successive five-second periods) and told to record the exact position of a "walker" (who has previously been instructed to maintain a steady pace). This is done for three different walking rates and the resulting distance vs time data is plotted (see Fig. 1 for a typical plot using student data). Although the procedure is simple—almost naive—it does produce these results:

(a) Students grasp the meaning of a straight-line graph. In later work, they refer back to this result as one whose meaning is clearly understood.

(b) It introduces the general notion of speed (v) as a distance interval ($\triangle S$) covered in a given time interval ($\triangle t$).

(c) It relates this speed to the slope of the line, recognized as a special case of the straight-line equation, $y = mx + b$, with zero the y intercept. Here, $S = (\text{slope}) \, t$, and since the slope is given by $\triangle S / \triangle t$, which is the constant[1] speed v_c, $S = v_c \, t$.

(d) It leads readily to the idea of distance as the area under the speed-time graph. In Fig. 2, the distance at any time t is given by the area of the physical rectangle having sides v_c and t. Thus, $S = v_c \, t$.

---

[1] The subscript c is used to distinguish the special cases of constant speed (v_c) and constant acceleration (a_c).

2. To introduce the notion of changing motion, subsequent "walkers" are instructed to either slow down or speed up, and the resulting graphs are examined qualitatively.

3. We study the quantitative concept of acceleration via free fall and the PSSC recording timer. From the graph of distance vs time (Fig. 3), values of the

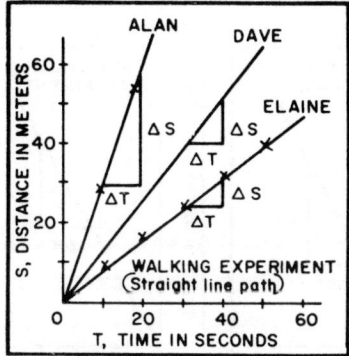

Fig.1. Distance vs time for three different walking rates. Speed in each time interval computed by $\triangle S/\triangle t$.

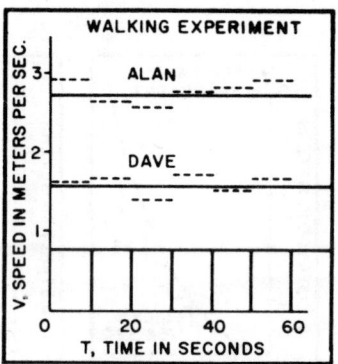

Fig.2. Speed vs time computed from data of Fig. 1. Area under v vs t graph gives distance at time t.

average speed in successive time intervals are obtained. At this point, we spend considerable time on the idea of instantaneous speed as the limiting value of average speed: $\triangle S/\triangle t$. Graphical values of the ratio $\triangle S/\triangle t$ are computed, starting with a long time interval and then continuing for successively smaller time intervals. We find this to be of considerable value in conveying the concept of instantaneous speed.

> **NOTE:** The S vs t curve is recognized as an example of second-power variation. Earlier in the course, this type of variation is introduced through an "experiment" whereby the area of a circle is measured as a function of the radius. After some discussion of the parabola, distance is plotted as a function of time squared (Fig. 4), and the slope of the resulting line is examined, its value computed, and dimensions noted to be those of acceleration.

4. Next, the speed values obtained from the S vs t graph are plotted as a function of time (Fig. 5). Students, recognizing this again as a straight line through the origin, immediately write v = (slope) t. It is not difficult to elicit also information that:

(a) The slope $\triangle v/\triangle t$ is the acceleration, by definition, and therefore, $v = a\,t$.

(b) The acceleration is constant, so $v = a_c\,t$.

(c) The distance can be computed as the area of the physical triangle, $S = \frac{1}{2}\,v\,t$, and since $a_c = v/t$, $S = \frac{1}{2}\,a_c\,t^2$.

(d) The value of the acceleration from the slope of this line is twice the value of the slope of the S vs t^2 graph.

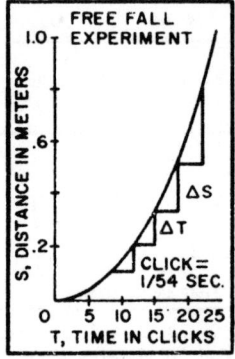

Fig. 3. Distance vs time for freely falling object. Data from tape of recording timer. Time unit (click) is period of timer, here, 1/54 second.

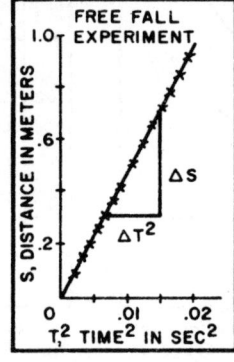

Fig. 4. Distance vs time squared, suggested by curve of Fig. 3. Straight line through origin gives S = (slope) t^2.

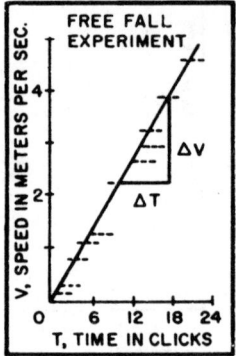

Fig. 5. Speed vs time computed from data of Fig. 3. Straight line through origin gives v = (slope) t. Slope is acceleration, by definition.

5. Keeping in mind that, in the case above, the object started from rest, we then consider the effect of imparting an initial speed, v_0 (Fig. 6). Here, of course, graphing v vs t yields a straight line again, but not now through the origin. The straight line equation now gives the more general expression, $v = v_0 + a_c\,t$. Also, under the v vs t graph, the area can now be computed by adding the areas of the physical triangle and rectangle, giving

$$S = v_0\,t + \frac{1}{2}\,a_c\,t^2.$$

NOTE: Other familiar expressions, such as $v = \frac{1}{2}\,(v_0 + t)$ and $S = \overline{v}\,t$, are obtained directly from the graphs. Some small amount of algebra will then deduce the final expression, $2\,a_c\,S = v^2 - v_0^2$.

VARIATION: We have used an interesting variation on free fall by using electronic pulsed flash (such as General Radio's Strobotac) and obtaining a Polaroid picture of the descent. Other timing devices like a metronome and seconds pendulum are used in observing other motions, such as the translation of a wheel rolling down a plane.

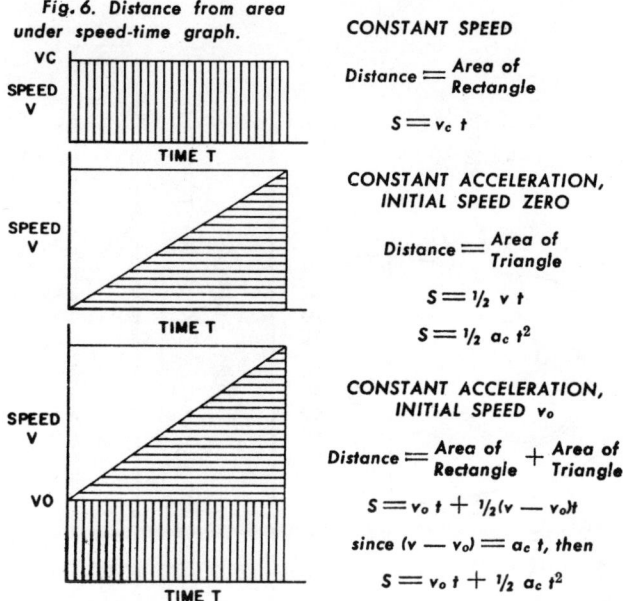

Fig. 6. Distance from area under speed-time graph.

CONSTANT SPEED

$$\text{Distance} = \frac{\text{Area of}}{\text{Rectangle}}$$

$$S = v_c\, t$$

CONSTANT ACCELERATION, INITIAL SPEED ZERO

$$\text{Distance} = \frac{\text{Area of}}{\text{Triangle}}$$

$$S = \tfrac{1}{2}\, v\, t$$

$$S = \tfrac{1}{2}\, a_c\, t^2$$

CONSTANT ACCELERATION, INITIAL SPEED v_0

$$\text{Distance} = \frac{\text{Area of}}{\text{Rectangle}} + \frac{\text{Area of}}{\text{Triangle}}$$

$$S = v_0\, t + \tfrac{1}{2}(v - v_0)t$$

$$\text{since } (v - v_0) = a_c\, t,\text{ then}$$

$$S = v_0\, t + \tfrac{1}{2}\, a_c\, t^2$$

We feel the approach works out well. Follow-up problems using the equations for motion with constant acceleration indicate that the material has been grasped well. More important than that: We have many indications that the students do acquire, through our development, some ability to collect, analyze, and interpret data meaningfully.

DEMONSTRATING THE INVERSE SQUARE RELATION

by Sister Mary Samuel

Siena Heights College, Adrian, Michigan

The presentation of the inverse-square law provides an opportunity to teach a variety of concepts and techniques using a single demonstration.

Materials

Solar cell (Bell Laboratories' *Sun to Sound* kit type)
DC power source

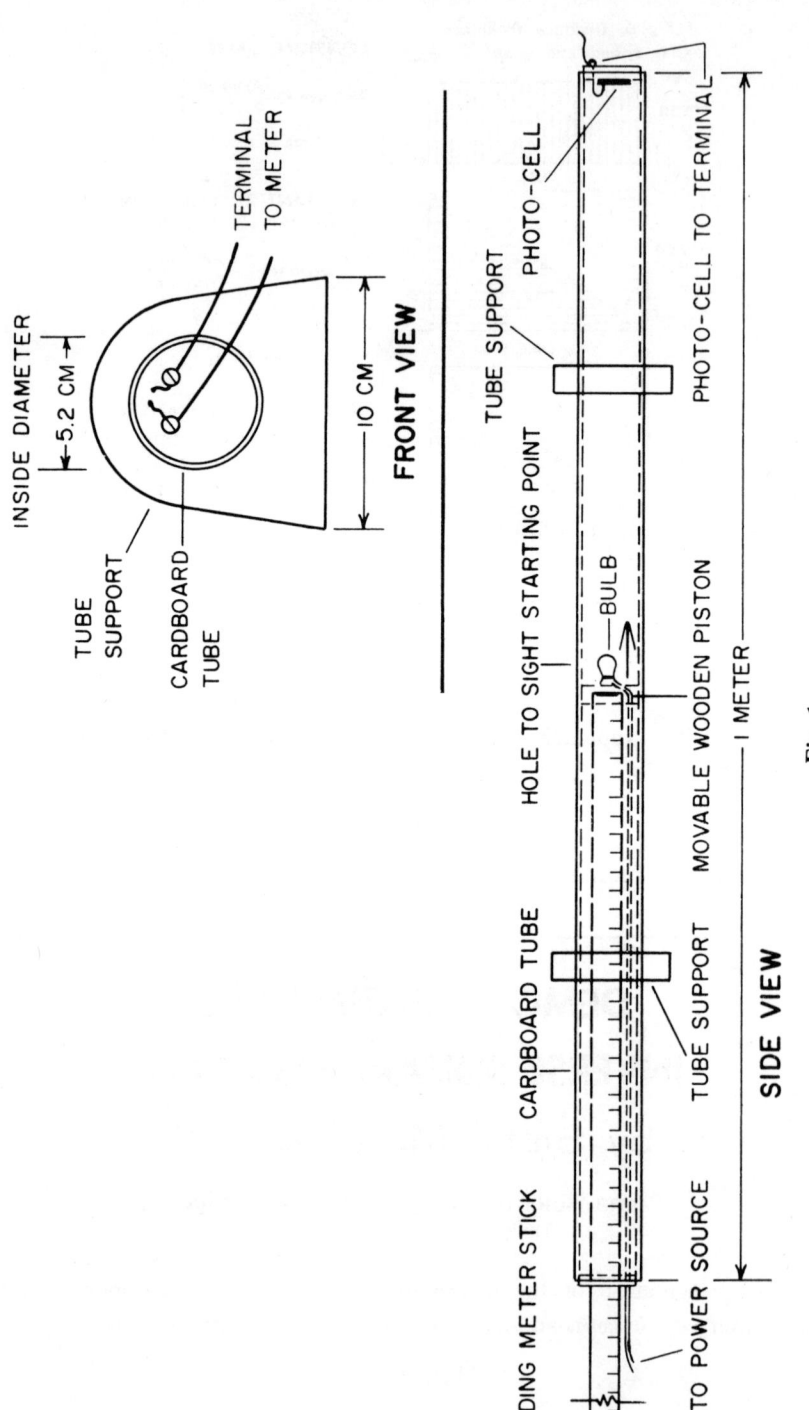

INSIDE DIAMETER

5.2 CM

TERMINAL TO METER

10 CM

TUBE SUPPORT

CARDBOARD TUBE

FRONT VIEW

TUBE SUPPORT

PHOTO-CELL

HOLE TO SIGHT STARTING POINT

CARDBOARD TUBE

BULB

MOVABLE WOODEN PISTON

TUBE SUPPORT

PHOTO-CELL TO TERMINAL

1 METER

SLIDING METER STICK

TO POWER SOURCE

SIDE VIEW

Fig. 1

120-volt, 6-watt bulb with socket holder (Sylvania)
Mailing tube (approximately 1 meter)
Meter stick
Microammeter

A drawing of the apparatus is shown in Fig. 1.

NOTE: I find it better to make the apparatus as permanent as possible, so that it will be ready for use each year. For the mailing tube, a cylindrical cardboard tube, such as those found in rolls of colored bulletin board paper, is ideal. Wooden supports and a painted exterior are not essential, but give a finished "professional" look to the apparatus.

Predemonstration Briefing

Distribute to each student a mimeographed copy of the diagram in Fig. 2. The accompanying presentation might be worded as follows: "Imagine several concentric spheres of radii r_1, r_2, r_3, etc., respectively. Suppose that an explosion occurs in the center of the smaller sphere. Particles would spread out in all directions, striking the curved surfaces.

"From each sphere let us mentally cut out a patch of curved surface of 1 square centimeter in area, and see what relationship we can find between the number of particles passing through the patch holes of the various spheres and the radii of the spheres."

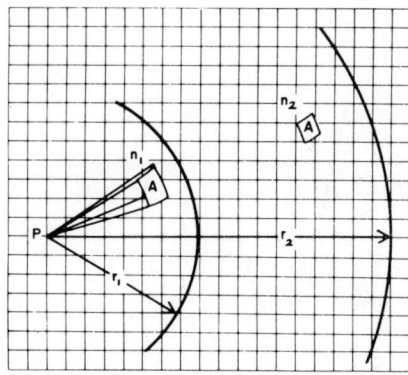

Fig. 2

P = *The number of particles available at Point O.*
A = *The area of patch on Sphere 1.*
n_1 = *The number of particles intercepted by A on Sphere 1.*
n_2 = *The number of particles intercepted by A on Sphere 2.*

Procedure

NOTE: Three or four students can carry out the mechanical work, such as reading the ammeter, recording data on the board, etc. Small groups find it exciting to gather around the demonstration table informally and record readings directly.

Letting the solar cell represent the patches, withdraw the light and meter stick from the tube at intervals of approximately 5 centimeters at a time. While the data is being gathered (Table 1), the students will notice that the number of microamperes produced by the solar cell is somehow related to the distance of the light from the cell.

Table 1: A Typical Set of Data

Intensity (microamps)	Distance (centimeters)
.50	52
.70	47

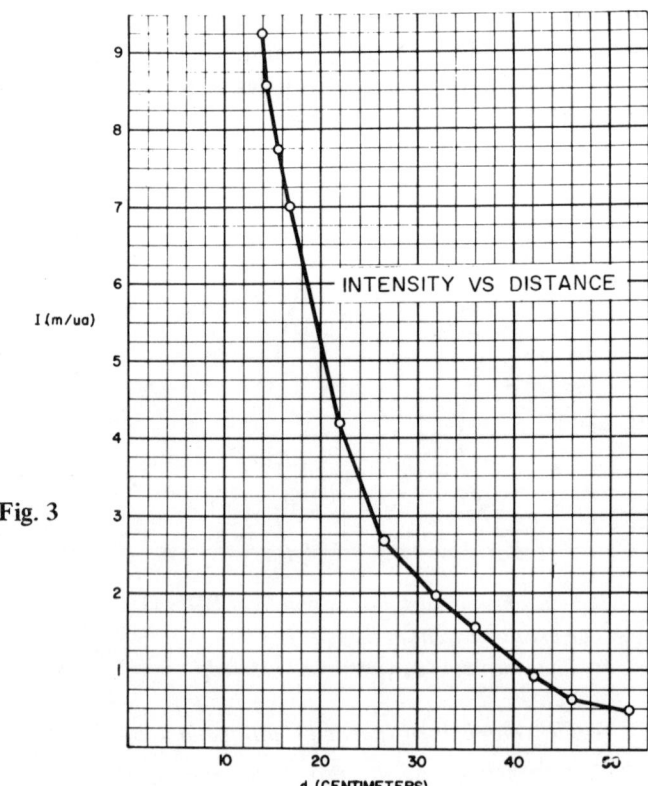

Fig. 3

.85	42
1.20	37
1.70	32
2.60	27
4.15	22
7.00	17
7.80	16
8.45	15
9.35	14

Upon examination of the data, the students will see that as the distance increases by small increments, the intensity decreases rapidly. This observation helps to develop the students' concept of an inverse relationship. At this point, graphs of I vs d (intensity versus distance) and I vs $1/d^2$ should be made (Figs. 3 and 4).

Fig. 4

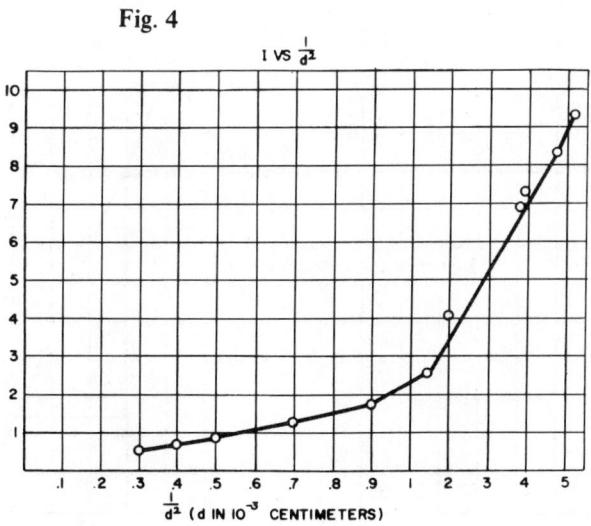

$I \text{ VS } \frac{1}{d^2}$

$\frac{1}{d^2}$ (d IN 10^{-3} CENTIMETERS)

The question arises, "Why plot $1/d^2$ and not $1/d$, or some other power?" One can determine whether the physical data follows a power law relationship of the form $y = Ax^b$ by plotting log I vs. log d (Table 2). If a straight line results, the slope "b" will supply us with the power.

Table 2: Logarithms of Data

Log I	Log d
−.30	1.70
−.16	1.67
−.07	1.62

— .08	1.57
.23	1.50
.42	1.43
.62	1.34
.84	1.23
.90	1.20
.93	1.18
.97	1.15

The slope (b) as determined from the resulting graph (Fig. 5) is −2.48, which implies that $I \propto d^{-2}$. Note that some of the points do not follow the inverse-square law since the law applies to *point sources* ideally. Moreover, the solar cell (our square patch) has a plane surface rather than a curved one. Nevertheless, the demonstration is a good one to show that data approximate the inverse-square law except when the light is quite close to the detector.

Any measurable quantity coming from a point source and traveling in

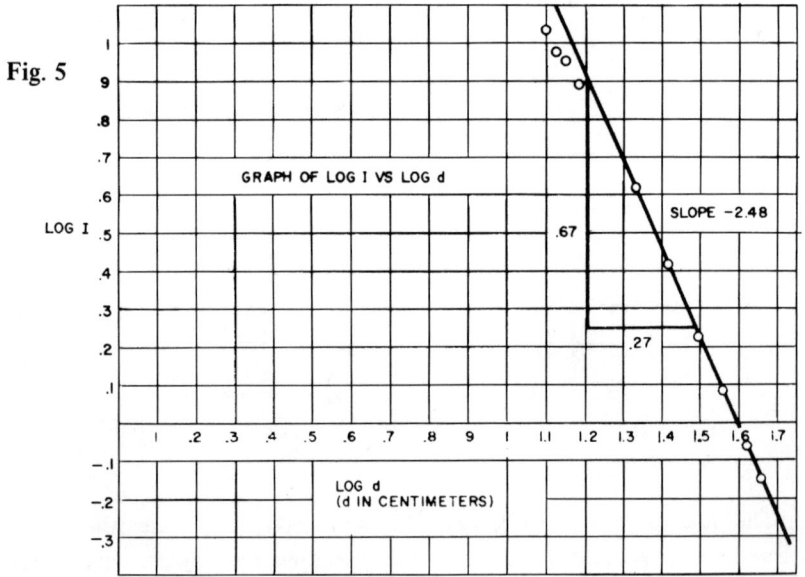

Fig. 5

straight lines will obey the inverse-square law. Actually, the *law* is proven by experiment, while the inverse-square relation can be proven mathematically. The fact that light does follow an inverse-square relation helps the student to realize that the light must be traveling in straight lines.

EXPLORING RELATIVE TEMPERATURES

by Donald P. La Salle

Avon Public Schools, Avon, Connecticut

How good is a measure? The high school laboratory exercise described here, which explores relative temperatures by calibrating ungraduated thermometers, is designed both to arouse the interest of the students and to create an appreciation for the built-in limitations of any measurement. Furthermore, it gives them an opportunity to discover some of these restrictions first-hand by utilizing rather simple equipment.

For a class of 30, the following equipment is needed:

15 ungraduated thermometers, approximately 25 cm in length (available from most scientific houses at about $1.50 each). Household thermometers, removed from their backing, might work reasonably well.

6 beakers (1,000 ml are satisfactory) for holding ice cubes or crushed ice.

6 Bunsen burners (or hot plates).

15 grease marking pencils.

15 shirt cardboards or large white construction paper.

Procedure

To prepare for the exercise, challenge the class by asking if anyone can read the temperature in the room by using one of the uncalibrated thermometers. You will get the obvious answer that it can't be done, which opens the whole problem for discussion. I find this to be an effective means of generating ideas: I ask each student to submit a brief outline, for the next lesson, on how he or she thinks the thermometers can be adjusted to give a reasonably accurate reading. No matter what technique you use, do not give an answer outright.

Invariably, there will be suggestions to use fixed points of some medium, usually culminating in the proposal that the freezing point and boiling point of water could be used. There is no difficulty in getting agreement that these limits will provide a convenient and accurate range. At this point, the problem of devising a method of calibrating begins.

Divide the class into groups of two or three, and permit each group to start calibrating from either end of the scale. Thus, half may attempt to record the freezing point with the buckets of ice, and the other half may concentrate on the boiling point.

When this has been accomplished, lay the thermometers out on a shirt board or construction paper and trace them. When the points experimentally arrived at have been included, the class is ready to figure out a way in which the space between the two points can be graduated. Again, this should be left to the ingenuity of the students, with the only requirement being that they represent both the Centigrade and Fahrenheit scale, preferably one on each side of the thermometer, as shown in Fig. 1. You will probably be surprised at the wide variations in methodology achieved by the individuals in the class.

NOTE 1: The limitation to Fahrenheit and Centigrade is one of convenience. Certainly, a student might decide to calibrate the thermometer in "hypograyphs," using a base 7 scale. Something like this would make the point that all our measurements are manmade.

NOTE 2: It is helpful, of course, to start with the Centigrade side by measuring the distance between zero and 100 degrees on the basis of some length divisible by 10. If this is impractical, then use a system which halves the total length. Further division into 20 readings of every 5 degrees will make the relationship between the two scales more evident, when the equivalent of 9 degrees is used on the Fahrenheit side. (See note to Fig. 1.)

When the students describe their conclusions, these questions should be answered:

Are you measuring heat or temperature?
Is there any difference between the two terms?

NOTE: Some may feel it is unwise to bring in the idea of heat at this point, as it may obfuscate the idea of relative temperatures. Heat vs temperature, of course, involves questions of major importance; but perhaps not here.

Are there any limits to the degree of fineness of your calibrations?
Are the only limits imposed those of your measuring devices?

As a result of the exercise, the students will undoubtedly have many questions of their own. Common ones that come up are:

How do we know when we have reached the end point of freezing when using water and ice?
Can an ice cube be colder than 32° Fahrenheit or zero Centigrade?
What is temperature really a measurement of?
What is heat?

Each of these questions can provide enough discussion for a lesson in

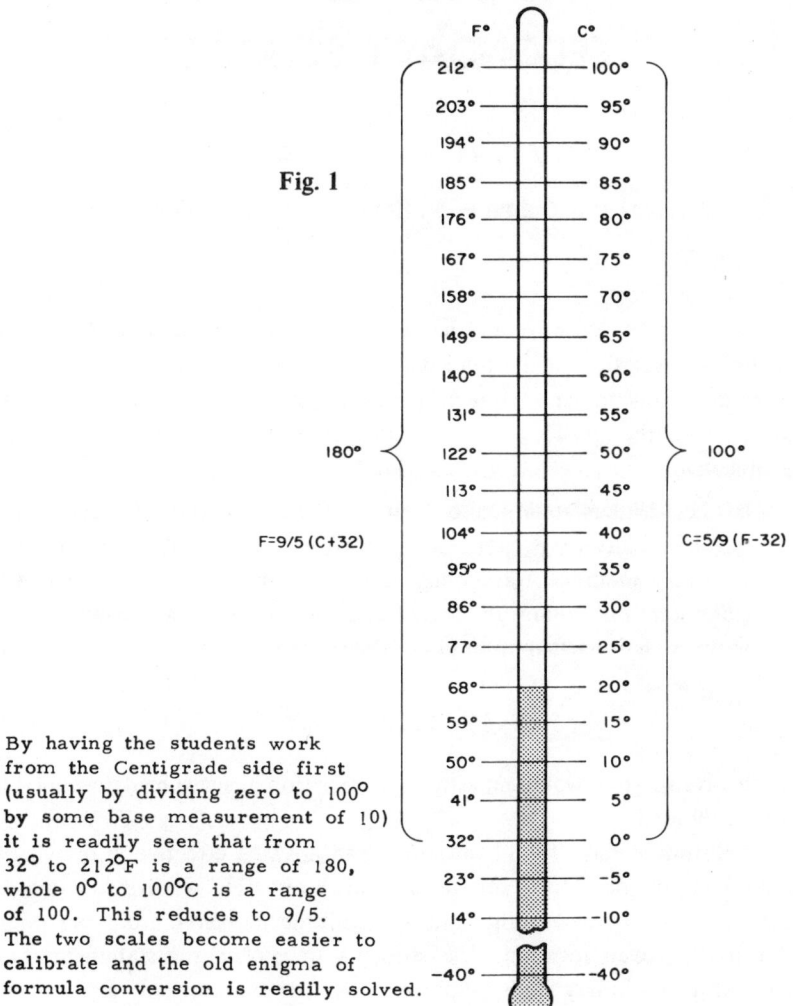

Fig. 1

180° $\Big\{$

F=9/5 (C+32)

100° $\Big\}$

C=5/9 (F-32)

By having the students work from the Centigrade side first (usually by dividing zero to 100° by some base measurement of 10) it is readily seen that from 32° to 212°F is a range of 180, whole 0° to 100°C is a range of 100. This reduces to 9/5. The two scales become easier to calibrate and the old enigma of formula conversion is readily solved.

itself; you can explore each of them as fully as time permits. You may find, for example, that it will come as a surprise to many students to know that ice can get as cold as most things; they have long associated ice with the freezing point of water.

With the groundwork thus laid, you can introduce the most challenging of all questions for discussion: How cold can anything get, and what are the limitations, if any, of man's ability to measure these depths? This could lead, of course, into discussions of absolute zero.

TEACHING VECTORS IN SECONDARY SCHOOLS

by William Naison

Formerly Jamaica High School, Jamaica, New York

One overall objective in the teaching of vectors should be to develop students' appreciation of the interrelationships between physical and mathematical models of phenomena. Though there are opportunities to present this idea in other parts of the physics course, the subject of vectors is particularly suited to this objective.

> NOTE: Understanding and feeling of the processes involved in science undertakings cannot be created without use of some kind of discovery method of teaching. Thus, following is a combination of classroom and laboratory activities programmed to stimulate maximum student participation in the time available.

I. The Position Vector

Problem: How would an astronaut's position in space be determined by an observer on earth?

Solution: I have found questions and answers effective in drawing from the students the need for a reference point, a direction relative to a coordinate system at the reference point, and a magnitude measured from the reference point to the given location. Ask students to explain the statement that the astronaut is stationary and inquire whether the observer on earth, given this statement, would measure the same information the next day.

A further help in developing this concept is having students make a graphical model representing the physical data used to locate the astronaut. Suggest the use of an arrow and the assignment of a name (position vector) to the line that most pupils will draw.

II. Introduction to the Displacement Vector

Problem: Give students the astronaut's position in terms of his position vector relative to a given reference point on earth, the latter *assumed* stationary with respect to some selected reference point (a star, for example). Then change

40

the astronaut's position in some defined way. How can the observer on earth find the new location of the astronaut?

Solution: Ask students to give the minimum information needed to constitute a "defined way of changing location." With encouragement, they will correctly write the given information at the head of the position vector.

NOTE: I believe that the least confusion results if the change of location data is given with reference to the selected coordinate system on earth.

Show how a coordinate system can be moved parallel to itself and transferred from the reference point on the earth to the head of the position vector. Have students lay off the angle and magnitude of the defined motion from the head of the position vector.

A good follow-up is a series of problem exercises for the following cases: (1) Given two position vectors, find the displacement vector (motion vector) necessary to change from one position vector to the other. (2) Find the final position vector, given a series of displacements beginning from the origin.

NOTE: Make a distinction between the position and the displacement vectors. Position vectors begin at the origin and represent a location from the origin. Displacement vectors may begin anywhere and represent a motion. The appearance and the rules of operation of both vectors are otherwise the same.

III. Components of a Displacement Vector

Problem: Repeat the problem in section II with the following modifications. Introduce a series of displacements such as those labeled *1-2, 2-3,* and *3-4* in Fig. 1a. Have students lay off the given displacement vectors and draw the final position vector. Can a single displacement vector replace the three separate displacement vectors?

Solution: Students will have no difficulty locating the single displacement vector, the *resultant* (Fig. 1b). Ask them to provide other possible components, using other displacements to other points but having the same resultant (Fig. 1c).

IV. Physical Verification

Problem: Present students with the following situation: A position vector from a reference system on the earth locates a rocket ship relatively close to the sun. Assume that there is no relative motion between any of the three objects and that a displacement, measured from the earth's coordinate system, is to be applied by remote control to the rocket ship. Find the new position of the space ship, noting any additional assumptions made in calculations.

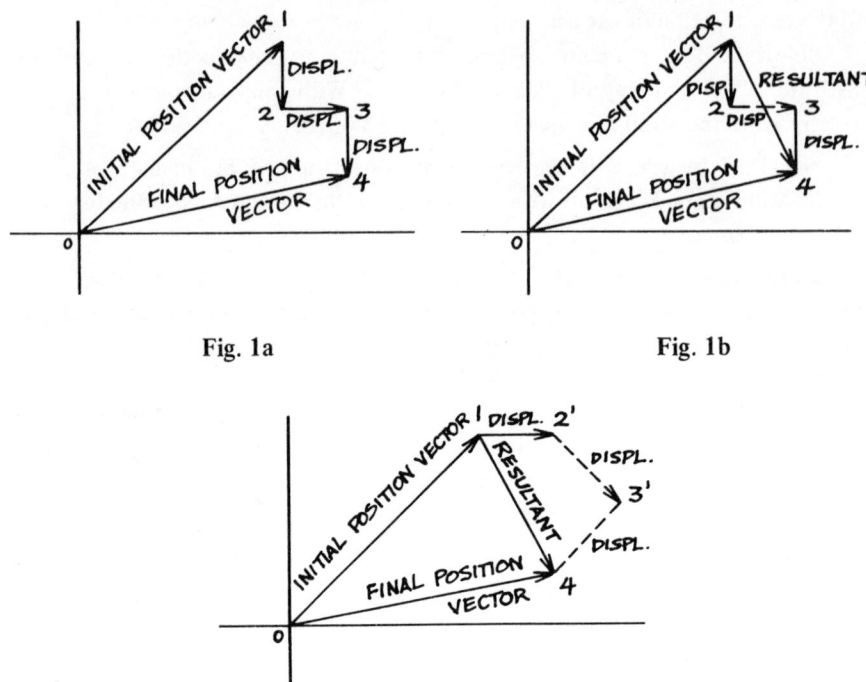

Fig. 1a   Fig. 1b

Fig. 1c

Solution: I have found that the best way for students to solve this problem is by means of a directed laboratory exercise, modifying the problem for simplicity, to find the displacement given the final position vector. Each student is provided with a hollow hemisphere made of some inexpensive material. Using the center of their hemisphere as the center of their coordinate system, students select any convenient point on the surface of the hemisphere as the position of the rocket ship. Each student then displaces the rocket ship to some new position, say about a quarter of the circumference. From the measured values of the two position vectors, he calculates the displacement on his mathematical model.

V. The Velocity Vector

NOTE: The scope in this sequence of activities does not include a series of lessons on motion, but the transformation of displacement vectors into velocity vectors requires some comment.

Ask students whether the velocity, *displacement per unit time,* can be represented in the same way as the simple displacement vector and if so, with what modification. Explain that time is a scalar and that velocity divided by time merely changes the magnitude of the displacement vector, not its direction.

To illustrate how velocity vectors might be utilized, a problem of the following kind might be used: A car driver involved in an accident is being sued. The opposing lawyer argues that the car driver, Abel, was negligent because he did not apply any corrective steering action. These facts are brought out at the trial: Car driver Abel was traveling at 60 mph along the roadway in the single lane of traffic when a cross-wind of 60 mph suddenly came up and drove his car to the other side of the road, causing a collision. Assuming no corrective action was taken by Abel, at what angle would he have crossed the road and at what speed?

VI. Physical Confirmation of Velocity Vectors

Problem: Two boys simultaneously kick a football from different directions. In which direction will the football go?

NOTE: I know of no simple quantitative demonstration being marketed designed to illustrate this problem. However, a simulated demonstration may be performed. The basic idea is to have each of two simultaneous motions, vertical and horizontal, respectively performed by two operators in the presence of an audible timing device such as a metronome.

Solution: The demonstration can be performed with a stiff piece of cardboard about 2 feet high and 1 foot wide, with a vertical slit through the center. As one operator moves the cardboard horizontally at a slow, predetermined rate, the other operator performs a constant vertical motion through the slit. For the vertical motion, use a rigid wire with a small but clearly visible sphere attached to the end of the wire facing the class. A crosshatch of lines, say 2 inches apart, are drawn on the cardboard, with the slit serving as one of the vertical lines. The crosshatch enables the class to observe the rate of movement and the operators to control it. Short horizontal lines on either side of the slit are drawn on the reverse side of the cardboard for the guidance of the operator. (See Fig. 2.)

NOTE: The operators may need practice to learn to produce a reasonably smooth motion. And it may be necessary to experiment with different rates of movement.

The actual operation of the demonstration should be complete in five or six seconds if the vertical movement is at the rate of 2 inches per second and the horizontal movement is at the rate of 4 inches per second. The initial reference point can be obtained by using a marker mounted on a ring stand, with the position of the latter also fixed in some way on the demonstration table. When time is called, the final position is held stationary to permit the class to note the vertical and horizontal displacements. Students compare the resultant physical displacement with the calculated displacement, expressed in unit time.

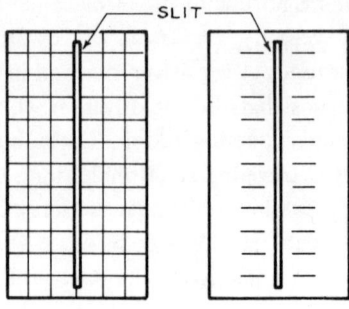

Fig. 2

VII. Application of Vectors to Forces

Problem: A small model of a conveniently weighted sign is supported by a spring balance. The problem is to support this sign by means of two supports on either side of the sign and some distance away—there are no vertical supports.

Solution: Have students solve this problem in a laboratory exercise. Ask them to obtain quantitative data by recording the magnitude and direction of the forces that will replace the single vertical force. Applying their mathematical model of a vector to these forces, they will verify its validity by calculating the resultant and comparing it with the original vertical force. In this way, they make various "discoveries" about the number of possible pairs that can be chosen as well as the number of configurations possible using more than two displacement forces. Students then try replacing each of a pair of components with two others, checking all steps by calculating the resultant of the components chosen.

VIII. The Resolution of Forces

NOTE: Here we find one of the weakest links in the teaching of vectors. This subject seems to mystify students. Following is an approach which has worked effectively for me.

Problem: Is it desirable to pull a sled horizontally? How is a horizontal force applied? Is there a more convenient way? How is it possible to apply a horizontal force by applying a force in some other direction? (The apparatus to demonstrate this is a staple of most high school physics departments.)

Solution: Students may suggest that the original force could be replaced by horizontal and vertical components. Demonstrate that no matter which pair of components are selected to replace the original force, their overall effect remains equivalent to that of the original force! The aim is to get students to understand that if a vertical force were present cancelling the vertical component of the original force, the horizontal component would become the resultant effective force. The source of the desired vertical force can easily be traced to the weight of the sled. An alternate approach is to inquire, "Which force added to the applied force would yield a resultant in the horizontal direction?" (See Fig. 3.)

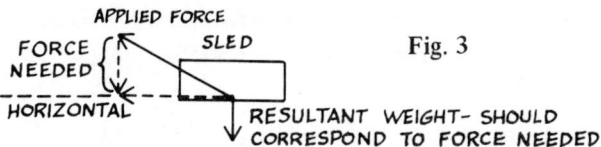

APPLIED FORCE

FORCE NEEDED

HORIZONTAL

SLED

RESULTANT WEIGHT— SHOULD CORRESPOND TO FORCE NEEDED

Fig. 3

To explain the quantitative aspects of this process, select a given weight for the sled and demonstrate that the resultant is horizontal only for certain selected applied forces—those with vertical components equal to the weight of the sled. Calculations can be checked by equilibrium methods. The applied force is balanced both by the weight of the sled and a horizontal force exerted in a direction opposite to the horizontal component. Friction can be introduced to explain any disparity between calculated and measured results.

IX. Independence of Forces Not at Right Angles to Each Other

Problem: Have students observe the velocity imparted to a freely falling body. After they have attributed the velocity to the force of gravity, ask them whether the vertical velocity would be affected if you simultaneously applied a force giving the ball a horizontal velocity.

Solution: Before referring to the mathematical model, it is necessary to convert the velocity vectors to displacement vectors and these, in turn, to position vectors in order to avoid extraneous problems. All this requires is having students confine their analysis to a one-second interval and using the displacement vector from the origin.

The analysis of the mathematical model is carried out by means of a comparison technique. Fig. 4a shows a vector corresponding to the average

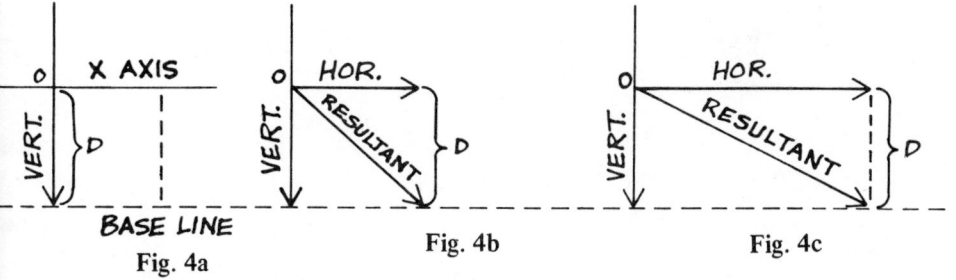

X AXIS
VERT.
D
BASE LINE
Fig. 4a

HOR.
VERT.
RESULTANT
D
Fig. 4b

HOR.
VERT.
RESULTANT
D
Fig. 4c

velocity of free fall. This vector is here regarded as the position vector of a freely falling object at the end of one second. Figs. 4a, 4b, and 4c show the resultant position vectors when different horizontal velocities have been imposed on the freely falling object. From the diagram, it is evident that the resultant position vectors are the *same* vertical distance away from the X axis as the freely falling position vector. The horizontal motion has no effect upon the vertical motion.

X. Component of a Vector in a Desired Position

Problem: Is the independence of two vectors due to the choice of horizontal and vertical directions or to the particular angle between them?

Solution: Using the same apparatus as in section IX, aim the apparatus at some angle to the horizontal and have students observe that the two balls no longer reach the ground at the same time. Refer to the mathematical model, Fig. 5, and compare the following vectors: vertical, the resultant of vertical and horizontal, the resultant of vertical and 30 degrees above the horizontal, the resultant of vertical and 60 degrees above the horizontal, and the resultant of vertical and 30 degrees below the horizontal. Students will see that the resultant position vectors are no longer in the same position with respect to the X axis. Ask them what the effect of each of these additional vectors is upon the vertical vector. They will be able to measure the effect by observing how far each resultant position vector is above or below the base line.

Next, ask students to suggest a simple construction that will enable them to find the effect they have just measured. If they have difficulty suggesting the projection of one vector upon another, a question may be framed in terms of Fig. 5. Draw Fig. 5b to scale and on the same diagram draw the projection of vector A upon vector V. Have students note the equivalence of the projection with the measured effect of vector A upon vector V. This process may be repeated for Figs. 5c, 5d, and 5e. When all of the diagrams are on a chart, rotation of the chart will show that if all of the relationships hold in some direction other than the horizontal, the same rules apply. There is nothing unique about the horizontal and vertical directions. If the projection of one vector upon another is zero, its effect upon that vector is zero.

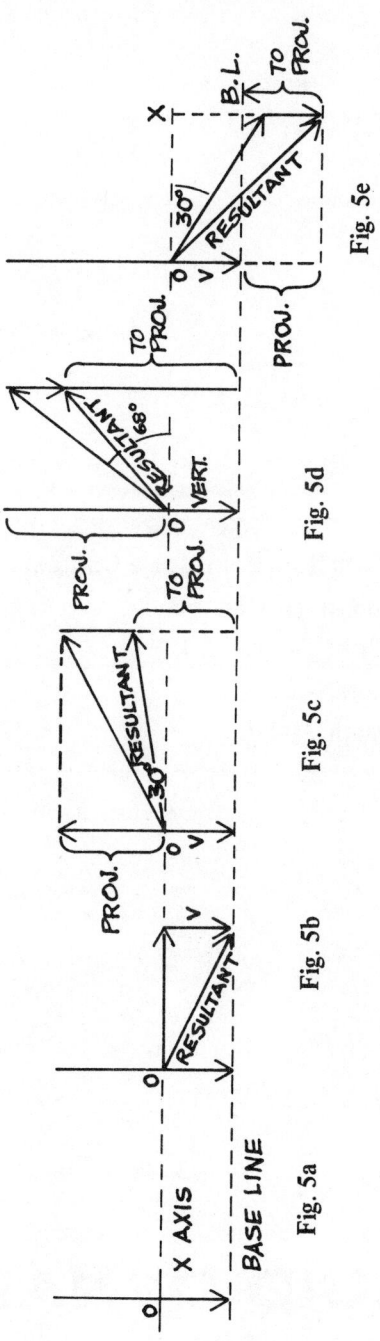

Fig. 5a

Fig. 5b

Fig. 5c

Fig. 5d

Fig. 5e

2

FLUID DYNAMICS

DETERMINING THE OVERFLOW POINT FOR A CONTAINER OF WATER

by Alfred F. Taylor

Ridge High School, Basking Ridge, New Jersey

This experiment is derived from a simple parlor trick, in which a glass is filled with water to the overflow point and a number of pennies are dropped into the glass without causing the water to overflow.

The Problem

Is it true that pennies can be added to a container which has reached the overflow point, or could more water be added as well as more objects such as pennies?

The approach: Some method must be devised to determine the maximum amount of water which could be placed into a given container before causing the water to overflow. The container must be filled with this amount of water. Attempts to drop other objects into the filled container will then be made.

The experiment: The maximum amount of water which a container might hold can be obtained by placing the container on a balance, and carefully weighing the amount of water which can be added until overflow is reached. For practical purposes, small beakers are suggested.

ASSUMPTIONS: The experimenter must assume that the size and shape of the container will have no effect upon the outcome of the experiment. This assumption can be tested by using different containers. The experimenter is forced to make several other assumptions; herein lies a considerable part of the project's value.

Once the combined weight of beaker and water has been determined, the beaker is again placed on the balance and filled with the predetermined weight of water *less that one last drop* which would cause the beaker to flow over. Now shot are carefully added until one shot causes the beaker to flow over. The shot are removed, dried, and added to a measured amount of water in a graduated cylinder. The rise in volume of the water determines the volume of the shot added.

49

Observations: (See following tables)

Table 1

Weight of beaker and water
at the overflow point

Trial number	1	2	3	4
Total weight (g)	43.00	43.05	43.00	43.05

Table 2

Calculation of vol-
ume of shot added
to the overflowing
beaker

Trial number	1	2	3
Total weight (beaker and water)	42.85	43.02	43.00
Number of shot added	26	13	18
Volume of shot (ml)	.47	.23	.32

Conclusions: From the data presented in Table 1, we make the assumption that when the total weight of beaker and water reaches 43.05 grams, any further additions of water would cause an overflow. From Table 2, we see that the total volume (or weight) of water can be increased by approximately 0.25 ml beyond this overflow point, assuming that the milliliters of shot can be converted directly into grams of water.

Thus we conclude that shot, and presumably pennies, can be added to a container full of water; whereas, if one had tried to add water to the container it would have overflowed.

A Second Problem

Why can one add more shot to a "completely" filled container without causing overflow, when the addition of more water would cause the overflow?

The approach: This problem requires more thought and research than the first one, and a closer observation of the experiment so as not to miss certain fleeting phenomena. For example, during the performance of the first part of the experiment, there will very likely be times when water dropped from an eyedropper onto the convex water surface will roll off the surface like a ball rolling downhill.

Some procedures to aid in understanding the nature of the phenomena might include greasing the lip of the beaker; adding drops of an electrolytic solution to the "filled" beaker rather than adding water; adding drops of soap water, alcohol, or oil to the beaker full of water.

The experiment: Using the same beaker, place in all but that last drop of water (for my beaker, the total weight of beaker and water is 43.00 grams maximum). Now add drops of saturated NaCl solution until overflow occurs, and record the new total weight (done in Table 3).

Using the same beaker, place in *less* than the maximum amount of water by approximately 0.5 ml (for my beaker, to a total weight of beaker and water of 42.50 grams). Now add drops of soap water, oil, or alcohol and record the weight when overflow occurs (done in Table 4).

Observations: (See following tables.)

Table 3

Addition of NaCl
solution to a "filled"
beaker

Trial number	1	2	3
Weight before adding NaCl	43.00	43.00	43.00
Weight after adding NaCl	43.35	43.25	43.30

Table 4

Addition of alcohol
to a "filled" beaker

Trial number	1	2	3
Weight before adding alcohol	42.50	42.50	42.50
Weight after adding alcohol	One drop of alcohol caused each beaker to overflow immediately.		

Conclusions: Again assuming that the beaker would weigh 30.05 grams when "completely" full, the data from Table 3 indicates that considerably more NaCl solution can be added than pure water, since approximately 0.25 ml of solution was added above the "capacity" of the beaker.

Why would this be so? Presumably one can pile water up in a beaker due to a combination of cohesive and adhesive forces existing in the system. When water is added to this pile, the drop of water must puncture the surface of the water in order to become an integral part of the water in the beaker. In so doing, one apparently has a situation very similar to the pricking of a blown-up balloon, except that where a balloon will immediately rupture, the water has a remarkable ability for mending its surface.

The drop of salt water, being highly ionic, must have a tendency to pull the water together as it touches the convex surface and passes through and/or contributes to the molecules making up the surface area. One might become involved here with a consideration of hydrogen bonding. That is, how does hydrogen bonding affect surface tension, and what is the relationship of this bonding to the rupture of the surface tension by drops of NaCl solution or pure water?

Assuming that we could always add water to the beaker (I used) until the total weight reached 43.05 grams, the overflow caused by the drop of alcohol (as indicated in Table 4) cannot be attributed to the addition of liquid volume to a "completely" filled container. We should be able to add 0.5 ml of alcohol, not just one drop (0.05 ml).

Following the same line of thinking concerning surface tension, adhesion, and cohesion, apparently the addition of nonpolar alcohol to the surface of the water so seriously disrupts the surface tension that the entire pile of water gives way. Thus, the alcohol has the opposite effect of the highly ionic NaCl solution.

NOTE: How does this help to explain the ability of shot to be added to a "completely" filled beaker?

The shot seemingly can pass through the surface of the water without breaking the surface tension. That is, the shot passes through the layer of water molecules without causing any disruption of the water bonds, even though it disrupts the water surface by causing a ripple. The NaCl solution pulls the water bonds in more tightly, and the alcohol either produces a complete surface of more weakly bonded alcohol or else just weakens the surface tension by interspersing alcohol molecules in with the water molecules.

From this, one might suggest that the water did not "wet" the shot very strongly, for this action would have necessitated a shifting of water bonds. And in this system where we have these bonds under maximum tension, this should have resulted in a rupture of the surface area.

Final discussion: There are other small revisions in the experimental procedures which deserve some thought and possible trial.

(1) It was suggested that the lip of the beaker be greased before adding the water. Although this has been done, the results have not been presented. Suffice it to say that they were just the opposite of what I had anticipated.

(2) One might devise a U-shaped apparatus, with one leg longer than the other. Then determine whether or not the U-tube could be more completely filled by adding water through the short leg or the long leg. A modified funnel or thistle tube might be used.

(3) Only one size beaker has been used, a 25 ml beaker. What effect does depth of container, or width of container, or use of a lipless container have upon the problem being investigated?

(4) What difference would it make if the shot were greased? How would glass beads work?

(5) A needle can be floated on the surface of water. Will a needle still float when the surface tension has apparently "reached its breaking point" in the completely filled beaker?

Such suggestions as these have not yet led to any experimentation. The work required is simple, as is the necessary apparatus. The experiments offer young students a chance to develop new problems, plan their own approach, collect data, and reach conclusions.

PASCAL'S LAW AND
A POP BOTTLE

by Kent V. Frank

Kurtz Junior High School, Des Moines, Iowa

In looking for a way to demonstrate Pascal's Law on the transmission of pressure in a fluid to my ninth grade science class, I came up with a trick I learned from a friend. I fill a pop bottle to within an inch of the top with water, strike the mouth of the bottle a sharp blow with the heel of my hand, and the bottom neatly falls out. (The pressure applied to the opening of the bottle is transmitted without loss to the bottom. The larger force resulting in the application of the same pressure to a larger area causes the bottom to break out.)

Fig. 1

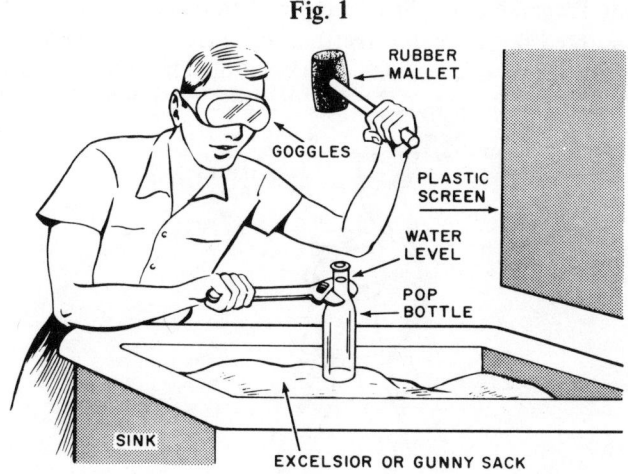

RUBBER MALLET

GOGGLES

PLASTIC SCREEN

WATER LEVEL

POP BOTTLE

SINK

EXCELSIOR OR GUNNY SACK

NOTE: Admittedly, the scientific principle here is momentarily obscured by the nature of the trick. I find, however, that the lesson stays in the minds of the students and that they appreciate the workings of hydraulic power after that.

To do the trick the way I described it takes a little practice, and could be quite painful to the hand. I therefore recommend a more civilized version of the demonstration, as shown in Fig. 1.

NOTE: Some may feel that assigning this demonstration at the ninth grade level is a little premature, since Pascal's Law is generally

covered in the physics course. At that level it has more meaning, since the mathematics can be indicated. I find, however, that it works quite well at the ninth grade level. Even if the mathematics is not understood, the students learn empirically that the principle operates.

THE CONSIDERATE
COKE BOTTLE

by Michael A. Iarrapino

Crosby High School, Waterbury, Connecticut

I use this to demonstrate atmospheric pressure at the start of that particular unit. Preparation is a little tricky and requires some patience (you may break two or three bottles before getting one that works). Drill a small hole about three-quarters of the way down a Coca-Cola bottle, or any pop bottle as shown in Fig. 1. (The Coke bottle is best because the green color helps conceal

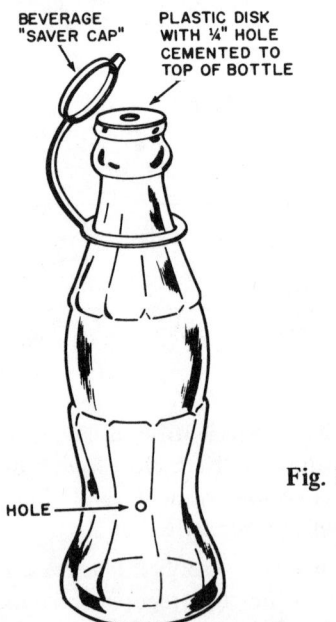

BEVERAGE
"SAVER CAP"

PLASTIC DISK
WITH ¼" HOLE
CEMENTED TO
TOP OF BOTTLE

HOLE

Fig. 1

the hole.) Build a small "dam" of putty around the area to be drilled. Fill with paste made of carborundum powder and kerosene. Drill (slowly, to prevent buildup of heat) with an electric drill having a hardened point. Reduce the mouth of the bottle to a hole about one-quarter of an inch in diameter by cementing in a small plastic disk with a hole bored in the center (the bottom of an ordinary plastic vial will fit nicely).

NOTE: A pre-prepared bottle of this type can be purchased from many magic dealers. I use the Coke bottle because it is recognized as authentic by the class. For the same reason, I avoid using bottles of metal or plastic, plus the fact that some substitutes might lack sufficient transparency for the full effect of the demonstration.

Fill the bottle with a brown liquid resembling Coca-Cola (don't use the real thing; the carbonation will spoil the effect). Fill with a funnel while blocking the hole on the side with your finger. Snap an ordinary "saver cap" on the bottle. Atmospheric pressure and the formation of a "Torcellian vacuum" will prevent the liquid from leaking out of the hole on the side.

I present the demonstration in a humorous vein. I pick up the bottle, secretly covering the hole with my finger. I remove the saver cap, looking around for a glass to pour the Coke into. As I do this, I "accidentally" allow the bottle to tip over so that the mouth is downward. The atmospheric pressure prevents the liquid from pouring out as surface tension does not permit air to enter through the reduced opening. Acting surprised, I pick up a glass tumbler and place it under the down-turned bottle mouth and permit the Coke to be poured into it by removing my finger from the hole in the side. I can control the flow by secretly covering or uncovering the hole on the side. I find this amuses the students and they become far more eager to study atmospheric pressure to find out the reason for the mystery.

MAKING A HYDROMETER

by Edward A. Terry

Science Teacher, Mobile County School System, Alabama

As the concluding activity in our study of density, I have each of my students make his own hydrometer which he uses in class to measure the relative density of several liquids, and then others of his own choosing at home. This learning experience gives him a working concept of relative density and an idea of its practical importance.

Materials

To make his own hydrometer, each student needs the following materials:
Plastic straw

3 or more small lead shot
Small rubber band or thin thread
Metric ruler
Soft lead pencil with a sharp point
Small test tube
Water glass

NOTE: The average class requires about two packages of plastic straws, one box of BB shot, and one package of thin rubber bands.

In addition, the class as a whole will require:

1 bottle of rubbing alcohol
1 box of table salt
1 box of copper sulfate (blue stone)

NOTE: The teacher should mix the copper sulfate with water and prepare the sodium chloride solution before classes.

Procedure

In presenting this exercise, I place written instructions on the blackboard, then carry out the instructions in a demonstration, having the students follow each step closely. (The directions might be written out and distributed to the students.)

1. Measure 5 cm from one end of the straw and make a bend at the 5-cm point.
2. Insert one small shot in this short end. Flatten this end, pushing the shot as near as possible to the bend.
3. Fold over at least 3 cm of the short end.
4. Insert two or three small shot in the long end and work them to the bend, being careful not to crush the tube.
5. Wind the rubber band closely above the shot.

NOTE: The straw should now be watertight and the shot firmly in place. If the steps have been correctly done, the straw will float in an upright position. (Usually, poorly prepared ones can be corrected.)

6. Place the straw in the glass of water and, with the straw floating in an upright position, mark the water line with the soft lead pencil. Check this mark carefully.
7. Remove the straw from the water and, with the fine-pointed lead pencil, mark a gauge of at least 2 cm above and below the water mark using milliliters as the gauge lines.

Testing the Hydrometer

Test the hydrometer first in alcohol, then, after washing it off, in the copper sulfate solution. The students will observe the fact that in neither solution does the hydrometer float near the water line. As a final test, place the hydrometer in the salt solution.

Following the instructions on the blackboard, which should list each step in the demonstration they have just observed, the students then make and test their own hydrometers. When these have been completed and tested, I encourage each student to continue his investigations at home on a suggested list of common liquids.

ATTENTION, ARCHIMEDES!

by Terrence P. Toepker

Xavier University, Cincinnati, Ohio

I introduce buoyant force by telling the class I could float the *Queen Mary* on a few gallons of water. Students, being what they are, almost always think this is impossible. Before demonstrating my statement, I try to find out *why* they think it's impossible, just to get the thinking processes started.

I tell the story of a clever sailor who bet that he could show how to float the *Queen Mary* on a few gallons of water. A friend immediately took the bet. Much to the surprise of the friend, the clever sailor drew some pictures which explained satisfactorily how the seemingly impossible feat could be done. The pictures are depicted in Figs. 1 through 4.

The sailor realized that the buoyant force exerted by a liquid is equal to the weight of the liquid displaced. Most people emphasize the word *weight;* the sailor concentrated on the word *displaced.* If the liquid is displaced (as in the case of an overflow experiment), it is not necessary to retain the displaced liquid. So, the sailor drew the pictures with the following explanation:

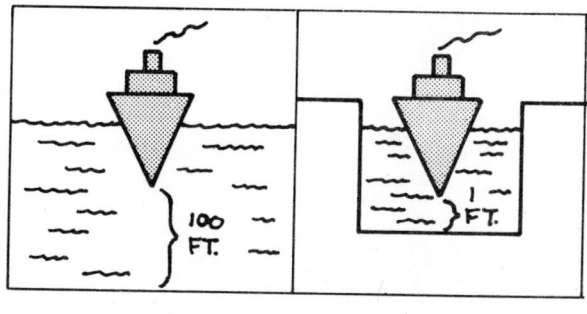

Fig. 1   Fig. 2

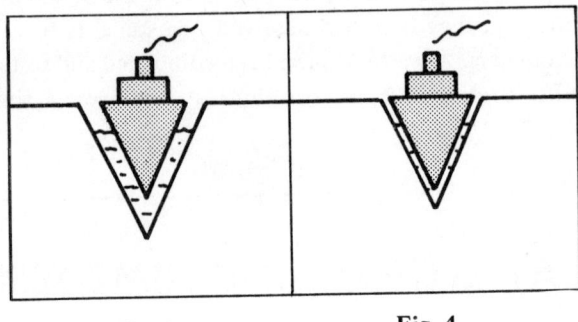

Fig. 3   Fig. 4

(1)   Fig. 1 shows a ship in fairly deep water.
(2)   Fig. 2 shows the same ship floating in a lock. So far as the ship is concerned, it floats just as well if it has 100 feet of water under it or 1 foot.
(3)   Fig. 3 shows the same ship in a special container where the walls slope in toward the middle. This reduces the amount of water needed to float the ship.
(4)   Fig. 4 shows the clincher. If the special container is made slightly larger than the ship, but following the contour of the ship, only a slight amount of water is necessary. Thus, it would be possible to float the *Queen Mary* on just a few gallons of water.

ILLUSTRATION: The principle can be easily illustrated with two nesting mixing bowls. One becomes the "ship" and the other the "lock" with the same contour.

3

OPTICS

MEASURING THE INDEX OF REFRACTION OF AIR USING A MICHELSON INTERFEROMETER

by James W. Crumley

Gold Beach Union High School, Gold Beach, Oregon

The idea for the following physics exercise developed indirectly from a student counter to my suggestion that we try to duplicate the Michelson-Morley experiment.[1] His comment was, "Why bother? We already know it will not work." At this point, our students knew that someone, somehow had measured the index of refraction of light in air, but they did not realize that the measurement could be done in a high school lab using a Michelson interferometer.

NOTE: Duplicating the experiment in our lab stimulated students' interest and gave them a sense of accomplishment from making careful measurements which are better than "reasonably close" to an accepted value.

Background

As most college-level physics texts provide a complete description of the Michelson interferometer, the explanation offered here is brief. Fig. 1 shows a schematic diagram of the instrument's components.

In operation, a light beam from an extended source of monochromatic light is divided by the beam splitter, bs, emerging as two perpendicular beams, each directed to a front-surfaced mirror—m_1 and m_2, respectively. These mirrors then reflect the beams along their original path to combine and are observed at

---

[1]The procedures described here are also developed in two other recent sources: "The Michelson Interferometer," *Physics—Advanced Topics Supplement,* Physical Science Study Committee, p. 158 (Boston: D. C. Heath & Co.); "Index of Refraction of Air," by Wallace A. Hilton in *The Physics Teacher,* April, 1968, p. 176. Mr. Crumley's work was done independently.

point O. The compensator, c, is a piece of glass the same thickness as the beam splitter and is necessary for use of the interferometer with white light, but unnecessary when using monochromatic light, as in this experiment.

A slight path difference will cause interference fringes if the mirrors are exactly perpendicular and if the light comes from an extended source. Moving one of the mirrors, say m_1, will cause the fringes to shift position; if one dark fringe replaces the adjacent dark fringe, the mirror has moved a distance of one-half wavelength of the light. Counting the fringes will enable the observer to determine how far the mirror has been moved. This method of measuring distances was used by Michelson.[2]

Procedures

To set up the experiment:
(1)   Place a vacuum chamber, vc, in the path of one light beam (Fig. 2).

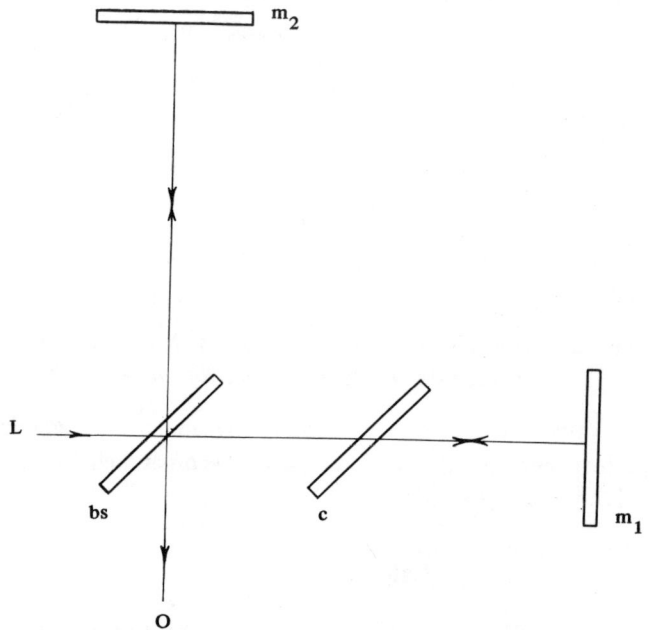

Fig. 1:   *L—light   source; 0—observer: bs—beam splitter; c—compensator; m_1—mirror; m_2—mirror. Although light is refracted when passing through glass, the light path is here represented as a straight line for simplicity of drawing.*

---

[2]Bernard Jaffe, *Michelson and the Speed of Light.* Anchor Books, Garden City, New York: Doubleday & Co., Inc., 1960.

Since light travels faster in a vacuum than in air, the effect is the same as shortening the distance, and the fringes move as air is allowed to enter the chamber.

(2) Place an index marker (needle stuck in a cork) in front of the beam splitter to serve as an aid in counting fringes.

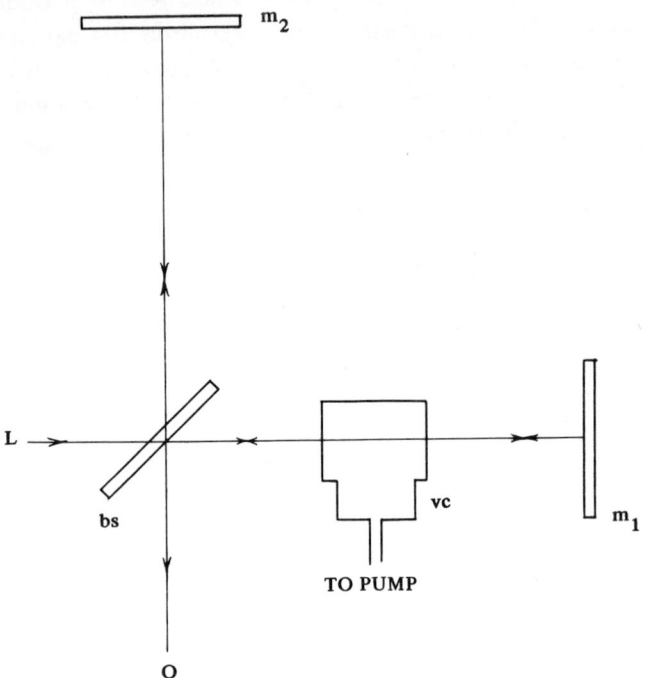

Fig. 2: *L–light source; 0–observer; bs–beam splitter; vc–vacuum chamber; m_1–mirror; m_2–mirror.*

AIR: For slow fringe movement, admit the air slowly. Practice makes perfect, but the rate of air admission becomes a challenge to students as countings are numerous.

Calculations

Calculating the index of refraction is rather simple. Students can go over the derivation of the equation with pencil in hand, verifying each step.

Let d be the length of the vacuum chamber. Thus the distance traveled by the light through the chamber is 2d. Also, let n be the index of refraction, t the time of travel through the chamber, and c the speed of light in a vacuum. Then

$$dn = tc \text{ and}$$

$2d(n_a - n_v) = (t_a - t_v)c$, where subscripts a and v represent air and vacuum respectively. Further, since

$(t_a - t_v) = N \dfrac{1}{f}$, where N is the number of fringes moving past the index and f is the frequency of the light source, and since $c/f = \lambda$, the wavelength of the light,

$$2d\,(n_a - n_v) = N\lambda$$
$$n_a - n_v = \frac{N\lambda}{2d}$$

$n_a = n_v + \dfrac{N\lambda}{2d}$, but n_v is taken to be exactly 1, so $n_a = 1 + \dfrac{N\lambda}{2d}$.

For our instrument, the vacuum chamber length, d, is 5.1 cm $\pm$ 0.5 mm, and thus is known within 1%, and the number of fringes with sodium light is about 50 $\pm$ 1/2 fringe, and thus is known within 1%. These two measurements, then, are compatible. Sample data of a typical run show N = 48 fringes:

$$n_a = 1 + \frac{48\,(5.893 \times 10^{-5}\text{ cm})}{2 \times 5.1\text{ cm}} = 1.00028 \pm 0.00001.$$

RESULTS: Since this method depends upon measuring the differ-ence between the index of refraction of air and of vacuum, and the difference is known within 1%, the actual index of refraction of air is known to about six significant figures. Few, if any, physics experiments for high school students produce such good results.

References

In addition to the work previously cited, these references should prove helpful in connection with the experiment:

Halliday, David, and Robert Resnick, *Physics.* New York: John Wiley and Sons, Inc., 1966.

Jenkins, Francis A., and Harvey E. White, *Fundamentals of Optics.* New York: McGraw-Hill Book Company, Inc., 1957.

Sears, Francis Weston, and Mark W. Zemansky, *University Physics,* 3rd ed. Reading, Mass.: Addison-Wesley Publishing Co., Inc., 1964.

THE AIR LENS

by George F. Smith

South Hadley High School, South Hadley, Massachusetts

Most physics students think of a double convex lens as being convergent and a double concave lens as being divergent. Quite frequently high school physics teachers will do several classroom demonstrations or laboratory experiments to reinforce this opinion.

The lens maker's formula, however, indicates that the characteristics of a lens are also dependent upon the relative index of refraction. The formula:

$$\frac{1}{f} = (n\text{-}1) \left(\frac{1}{R_1} + \frac{1}{R_2} \right)$$

Specifically, it indicates that the roles of the two lens shapes may be reversed by making a lens whose relative index is less than unity. To demonstrate this, you need the following materials:

Four watch glasses
Plastic electrical tape
Beaker
Masking tape
Dark and white paper
Point source of light (PSSC high-power light source)

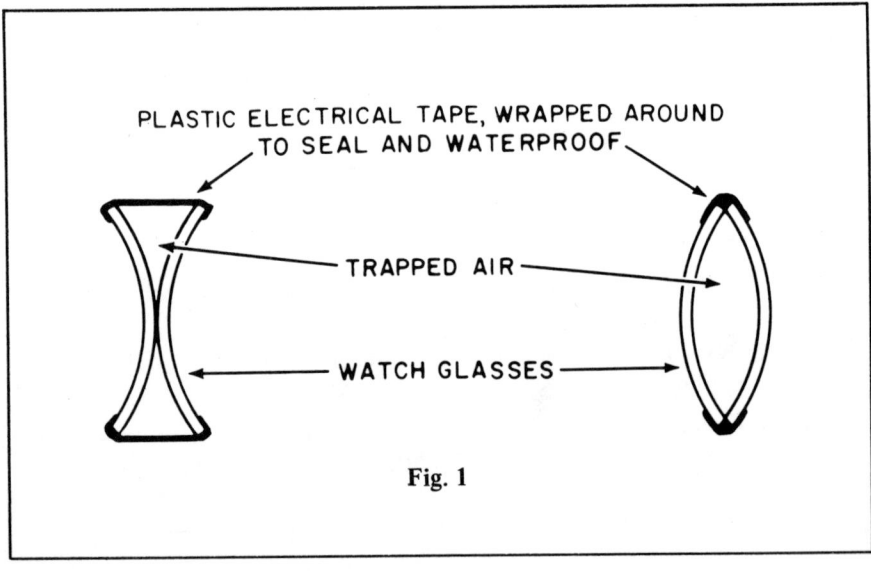

PLASTIC ELECTRICAL TAPE, WRAPPED AROUND TO SEAL AND WATERPROOF

TRAPPED AIR

WATCH GLASSES

Fig. 1

Matching pairs of watch glasses can be taped together to form a trapped air space of appropriate configuration to act as air lenses under water. Plastic electrical tape will seal them sufficiently well (Fig. 1). Cut two slits in the dark paper and tape with masking tape to one side of a beaker. Tape a piece of white paper to the other side of the beaker to act as a screen. Fill beaker with water and place air lens in water. Light from the illuminated slits (from the light source) on one side of the beaker can be seen on the screen on the other side (Fig. 2). The role of the lens shapes is clearly reversed; i.e., the double concave

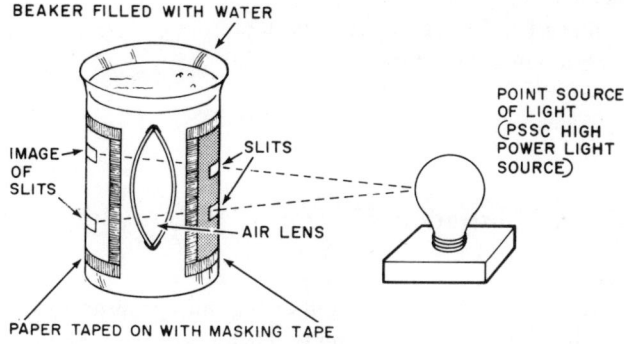

BEAKER FILLED WITH WATER

IMAGE→
OF
SLITS

SLITS

AIR LENS

POINT SOURCE
OF LIGHT
(PSSC HIGH
POWER LIGHT
SOURCE)

PAPER TAPED ON WITH MASKING TAPE

Fig. 2

ens is divergent, as shown in Fig. 2. Similarly, the double convex lens would be
onvergent.

DEMONSTRATING

OPTICAL DENSITY

by James O. Coats

Mandan Senior High School, Mandan, North Dakota

Optical density is often defined as the property of materials that tends to
nange the speed of light as it passes through a transparent material. This leads
o a question: What would be the effect, if any, on light passing through two
ntirely different materials if one could produce an optical density of a liquid to
natch the optical density of a solid, such as Pyrex?

I use two open-end demonstrations to show students the effect of optical
ensity on different materials.

**NOTE: As a part of the second demonstration, I can emphasize, by
analogy, the drainage patterns present in different types of soil.**

Materials Needed

$ZnCl_2$ solution, 61% by weight
Pyrex tubing, about 8 inches long

Reagent bottle
Rectangular tank or box, about 4" x 1" x 8" with drain
Pyrex chips of various, and uniform, sizes
Short section of rubber tubing
Pinch cock
Beaker

Demonstration #1—Optical Discontinuity

Fill the reagent bottle about ¾ full of the $ZnCl_2$ solution. Seal one end of the Pyrex tube by holding a finger over it. Then lower the tube into the reagent bottle containing the $ZnCl_2$ solution (Fig. 1).

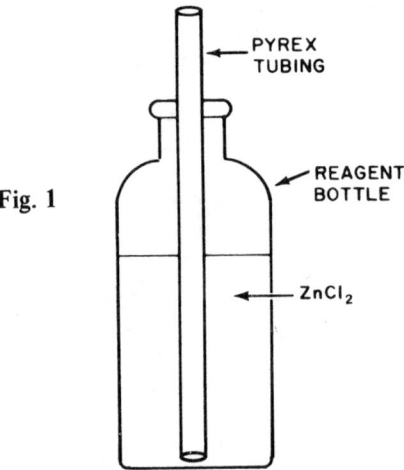

Fig. 1

First, allow the students to observe the tubing in the bottle of solution while air is trapped in the tubing and the pressure on the fluid is not great enough to force the solution into the tube. Remove the finger seal from the tube, which allows the solution to be forced into the tube. The submerged portion of the tube will seem to disappear.

Demonstration #2—Analyzing Drainage Patterns

Fill the rectangular tank about ¾ full of Pyrex chips. With the pinch cock closed, slowly add $ZnCl_2$ solution until the Pyrex chips are covered by the solution. The chips will appear to vanish as they are covered.

Open the pinch cock and allow the solution to drain from the tank in

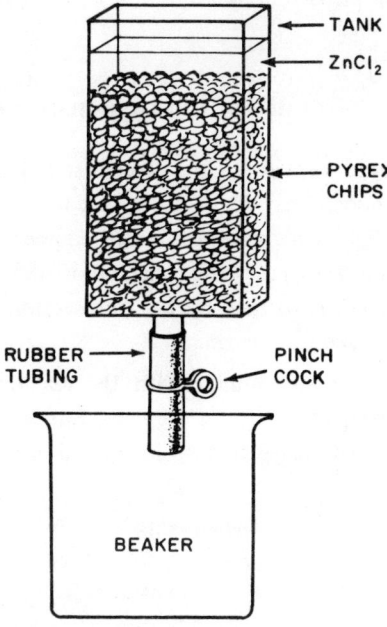

Fig. 2

he beaker (Fig. 2). Observe the drainage pattern as the solution drains from the
ank. Repeat the procedure and note that the drainage pattern does not repeat
tself.

By varying the size of the Pyrex chips, you may set up situations
nalogous to the different types of soil present on the earth's crust. Also, by
mploying, in different demonstrations, uniform Pyrex chips of a different size
or each demonstration, you may obtain more intricate results.

NOTE: The observations and data obtained from these demonstrations can lead to wide-open analysis and discussion. The many
offshoots presented to the students make both of them fine
open-end demonstrations or experiments.

A PARABOLIC MIRROR-REAL
IMAGE DEMONSTRATION

by Donna Parsons

Caldwell High School, Caldwell, Idaho

The following demonstration is a simple adaptation of a device suggested
n the *PSSC Teacher's Resource Book and Guide*. It has proven very effective in

my physics and general science classes and serves as an excellent introduction to the study of optics.

Background for the Demonstration

The teacher should explain to the students that although the image of ourselves we see in a mirror may seem real to us, technically, such an image is only virtual. Rays of light which form the image appear to come from *behind* the mirror while actually they are being reflected off the mirror's surface. A real image, on the other hand, is formed by the intersection of rays that really are coming from where they seem to originate.

NOTE: Students are all familiar with the real images formed by movie and slide projector lenses on screens. But few people have had the chance to see real images formed by curved mirrors.

Materials

To set up this real image demonstration, the following materials are needed: a parabolic or concave mirror, a strong light source, an object, and some means of holding the preceding three items in the proper position in relation to each other.

In the blueprint of the apparatus shown in Fig. 1, the mirror is an 18-inch

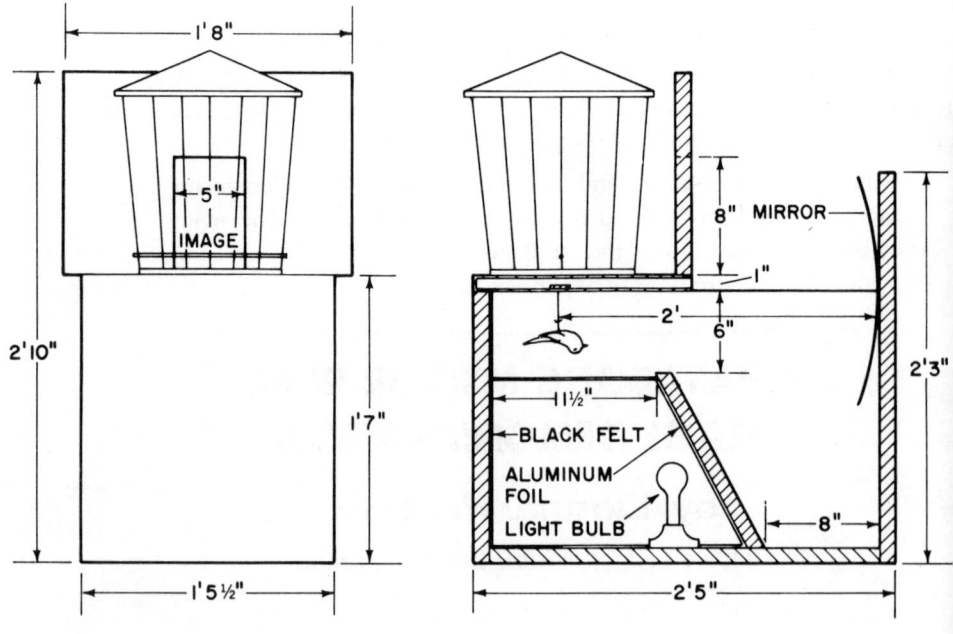

concave mirror with a focal length of 22 inches. The light is provided by a 200-watt bulb in a regular-sized base, and the object is an artificial bird. All parts are held together by a wooden box. Photos 1 through 5 depict the completed apparatus based upon the blueprint, removable parts of the finished apparatus, and a view of the image itself.

Apparatus Modifications

The materials used in the apparatus described here were selected largely because they were readily available. A smaller mirror than that shown here can be used, though the image might not be as bright as that from a larger one. The light bulb used might be stronger or weaker than the 200-watt bulb used in this model, and the dimensions of the box can be easily made larger or smaller than those of the model.

NOTE: It is essential, however, to be able to place the object at a distance equal to 2f, twice the focal length of the mirror, so that the image will be the same size as the object. Also, a slot or hook on which to suspend the object upside down must be provided, for real images are always inverted.

Although Fig. 1 and Photo 5 show a bird as the object, with the bird cage in front of the mirror to help the illusion, an evergreen bough can also be used effectively. Proper placing of the bough will make the image appear on a perch in the cage, or in the midst of the greenery. A flowers-vase combination or a dollar bill dangling in midair provide other variations.

1: *A view of all parts of paratus.*

Photo 2: *Picture of the position one must take to see the image. (The image will appear in the bird cage.)*

Photo 3: *View with the bird cage removed.*

Photo 5: *Picture of the i* *itself. (In actuality, the im* *is sharper than it appears h*

Photo 4: *View of the removable bird cage support.*

Using the Apparatus

The larger the ratio of diameter of the mirror to focal length and the brighter the light, the more noticeable the image will be. Lining the lower part of the box with foil will increase the reflection. Once the box frame is finished, the major problem is adjusting the mirror in the proper position to get an undistorted image. Then, since the image can be seen only when looking directly toward the mirror (Photo 2), the observer must be in the proper place.

NOTE: Arranging the apparatus so that students see it in operation as soon as they enter the classroom is most effective. Once they have all had a chance to notice how the image disappears with a change in their position, their interest is captured and motivation for learning is high.

The most practical use of this demonstration apparatus is in physics classes to demonstrate image formation, but it can also be profitably used in any physical science class to introduce a study of light, and in biology classes to illustrate the formation of images as related to the lenses of the eye.

VISUAL CONFIRMATION—
VIRTUAL IMAGE CHARACTERISTICS*

by John Robert Feather

Falls Church High School, Falls Church, Virginia

Physics students can be helped to understand the characteristics of virtual images using the method of Visual Confirmation. This method makes proof of virtual image characteristics realistic by being based on confirming visual observations. The observations are made while using a medium possessing both reflective and transmitting properties.

NOTE: One such medium—a two-way mirror or two-way medium—is described here, together with demonstration procedures for its use in the physics classroom.

Virtual Images and the Two-Way Mirror

Real image vs. virtual image: The light or any other electromagnetic radiation an object either emits or reflects, when collected properly, can produce an image. The type of image one observes depends upon whether the radiation from the object producing it is converging or diverging in space after being reflected or transmitted. There are two types of images—real and virtual.

REAL IMAGE: A real image is an image that can be reproduced on a screen because the image is produced by radiation that is converging after it has been reflected or transmitted.

A virtual image is the type of image a person can observe but which cannot be produced on a screen, because the radiation one collects to see it is diverging in space. The diverging expanding energy appears to be coming from an object deep within or behind the medium producing it. That observed object deep within is the virtual image whose characteristics are to be visually confirmed.

Simultaneous observation: The two-way mirror's reflective property allows a person to observe a virtual image produced by an object on the same side as the observer. Its transmitting property permits one to see objects behind the medium by looking through it. Using two similar objects, on opposite sides of the medium, one can observe both the reflected and transmitted effect

---

*This article is based on "Visual Confirmation—Virtual Images," a paper delivered by the author at the 1968 N.S.T.A. Convention in Washington, D. C.

simultaneously (Fig. 1). The simultaneous observation allows one to confirm visually, by comparison and parallax, the characteristics of the virtual image being studied.

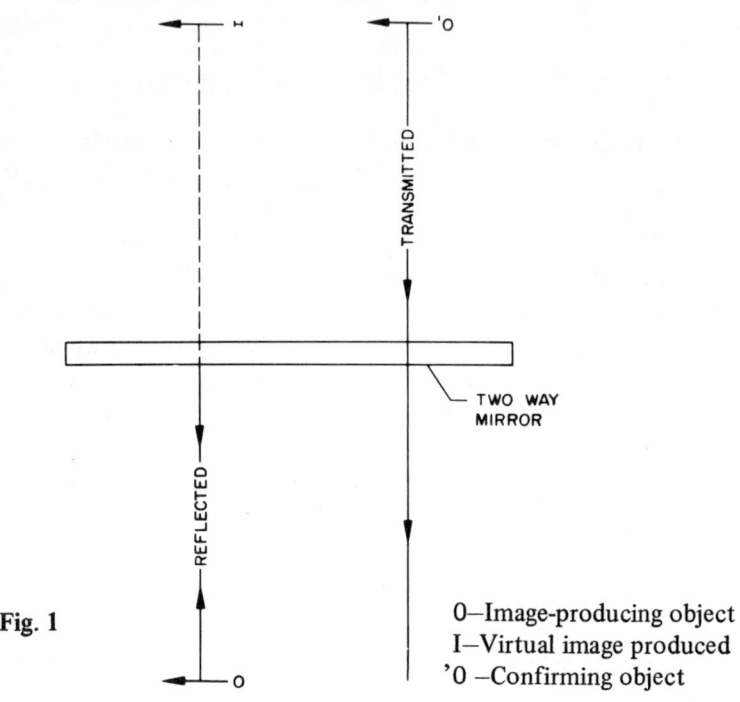

Fig. 1

0—Image-producing object
I—Virtual image produced
'0 —Confirming object

With this technique, the virtual images produced by the lens and concave or convex reflective surfaces can also be confirmed using the same two-way medium. The medium is placed between the observer and the lens or reflective surface and one observes the virtual image—that which is transmitted through the medium. The characteristics for the transmitted image produced by the lens or concave or convex reflective surfaces can be confirmed with a similar object on the observer's side by using the reflective property of the two-way medium.

Previous Methods of Studying Virtual Image

Before development of the visual confirmation method, there were two basic methods that could be used for virtual image study.

Recording lines of sight: One method consisted of positioning upon a piece of paper an image-producing reflective surface in front of an object. With the reflective surface perpendicular to the paper, a spot on the image is observed from different angles and the lines of sight recorded. Line segments representing the line of sight for the different observations are drawn from the paper to the reflective surface. The position of the reflective surface and the object must be recorded on the paper before removing them to continue.

The line segments of sightings for the same spot on the image are drawn till they intersect at a point, and the point of intersection is then accepted to represent where the observed spot on the virtual image has been. The position of the intersection point behind the reflective surface can be compared to the recorded position of the object, and one may accept that the image is the same distance behind the reflective medium as the object is in front of it.

PLANAR DRAWING: From recording observations for three or more spots on an image, a planar drawing can be made to aid in predicting the shape of the previously observed image. Using the drawing, one may predict that the image is the same size as the object, but reversed. The height of the object could best be checked by laying the object on its side and recording the extreme observations.

Using parallax: The second possible method a person studying virtual images could use predicts the virtual image location using parallax. Parallax is the apparent shifting of one stationary object with respect to another stationary object for observations at different locations.

One stationary object—the virtual image—is observed while at the same time one sights over and behind the reflective surface a second object. The second object behind the reflective surface is moved until it produces a visual effect of being a continuation of the observed virtual image, possessing no shifting for different observations' locations. The position of the second object is then accepted to represent the apparent location for the virtual image.

Advantages of confirmation method: Both of the earlier methods for studying virtual image characteristics require reflective surfaces called mirrors. A mirror with its metallic backing allows no energy, from an object, to fall behind it where the virtual image appears to be located, proving a virtual image could never be produced on a screen. The visual confirmation method, however, incorporates the use of a two-way medium having both reflective and transparent properties.

When using the two-way medium, it can be observed that energy from the object is both passing through the back and being reflected to the eye. The reflected energy produced the virtual image. The energy transmitted from the other side only allows one to see objects on the opposing side. (This is similar to the effect of looking out a window at night while in a well-illuminated room and seeing reflected and transmitted objects.)

Using the Visual Confirmation Method

Demonstration procedure: Following is a technique that the teacher may employ while using a two-way medium to confirm visually the characteristics for virtual images with his students.

(1) Support a planar two-way medium so that it is vertical to the area on

which the demonstration is being conducted.

(2) Place a very intense light source in front of the medium, making it appear to be a normal mirror.

(3) Secure two identically shaped objects—two 100 ml. graduates, for example—and place one of the objects on the intensely lighted student side of the medium.

(4) Point out the virtual image to the students.

(5) With the second object extended above what students think to be a normal mirror, use the parallax method to predict the location of the image.

> NOTE: The teacher, on the other side, can precisely position the object since he is able to see the transmitted energy from one object and the reflected energy from the second object.

(6) Instruct the students—now pleased with their understanding of virtual images—to watch that image. (They get a visual shock, as the image appears to remain after the object and intense light source are removed. It appears to remain because they are now seeing—by transmitted energy—the other object behind the two-way medium.)

> NOTE: Students are now visually prepared to confirm the characteristics for virtual images, having been shown the difference between reflected and transmitted energies using the planar two-way medium.

(7) Using both objects, on opposite sides of the medium, show that the image does appear to be the same distance behind the medium as the object is in front of the medium.

The one visual confirmed characteristic which amazes most students is that of the image being the same size as the object for any object distance from the mirror. The transmitted energy together with the reflected energy from similar correctly positioned objects appears to be one and the same, visually confirming virtual image characteristics.

> IMPORTANT: Make certain that the students see that for planar reflective surfaces the image is reversed compared to the object.

An exception: An exception to this image reversal which makes a good demonstration is produced by using two-way planar mediums at a right angle to each other. This arrangement creates three virtual images, and the location of the three possible images can be confirmed. At the same time, it will be seen that the middle image is not reversed compared to the object.

> SUGGESTION: In explaining the middle image, the object may be considered as a man moving his left hand. The middle virtual image also appears to move his left hand, while the other two images at the same time appear to be moving their right hands.

Virtual Images for Lenses and Concave or Convex Reflective Surfaces

After successfully using the preceding technique to confirm virtual image characteristics, students should be ready to study the virtual images for lenses and concave or convex reflective surfaces with the same two-way mirror.

Procedure:

(1) Place the lens between the image-forming object and the two-way mirror (Fig. 2). With a convex lens, the object is placed less than the focal length.

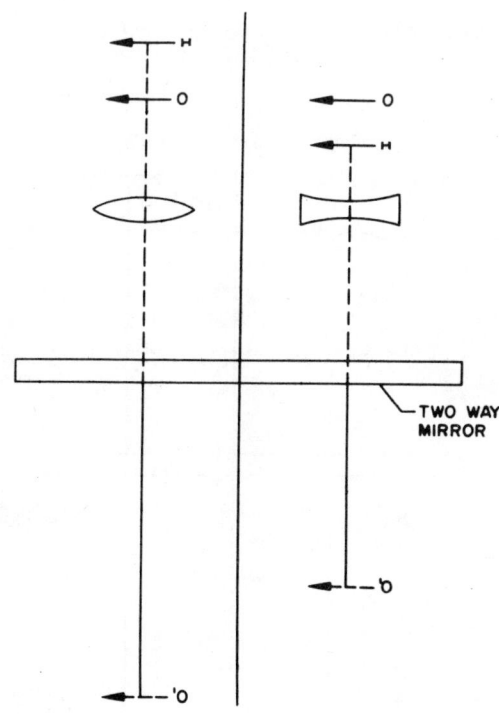

Fig. 2

0—Image-producing object
I—Virtual image produced
'0 —Confirming object

NOTE: To determine the focal length for a convex lens, allow energy from a distant object—beyond 100 yards—to produce a real image. The distance from the lens to the real image is considered the focal length.

With a concave lens, a small virtual image is obtained for any object location.

(2) Look through the two-way mirror at the object behind the lens; it is obvious that the observed image is really the virtual image produced by the lens.

(3) Using the reflection of an identical object, confirm the characteristics of the diverging transmitted energy that formed the virtual image. Compare the sizes of the two images.

NOTE: The comparison is more easily made when the objects used are identical in dimensions.

To confirm the virtual images with convex and concave reflective surfaces, using the two-way medium, place the object which produces the virtual image between the reflective surface and the two-way medium (Fig. 3).

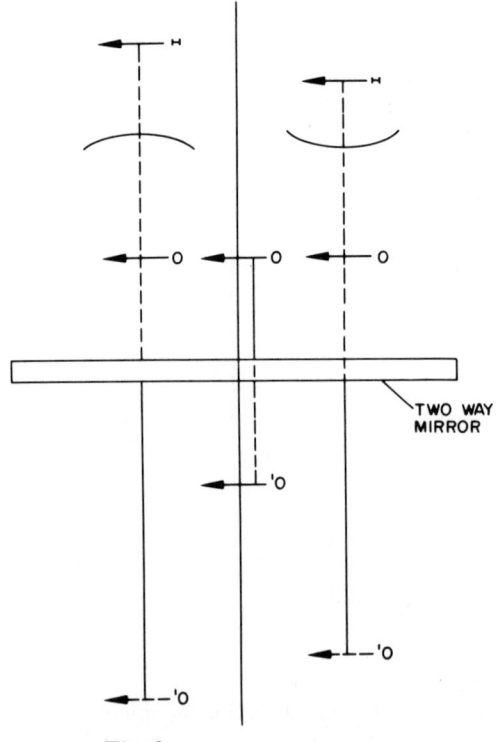

Fig. 3

0—Image-producing object
I—Virtual image produced
'0 —Confirming object

(4) Look through the two-way medium: The image and the image-producing object can both be observed at the same time.

(5) Using a second identical object, confirm the size and location for both sightings by superimposing the reflected image on the transmitted observation. A measuring device such as a meter stick can be used to obtain precise dimensional observations. The reflected stick is superimposed on the transmitted image and the magnification determined.

Microscopic Observations

The application of virtual image characteristic study can be extended to microscopic observations. An example of this application is the research technique with a microscope and mirror. The person using the microscope places the object to be studied between the objective lens and a two-way or regular-type mirror. This produces a virtual image which can be observed and allows observation of both sides of an object without having to turn the object over. The microscope can be focused either on the object or on the virtual image deep within the reflective surface. Both sides can be observed simply by raising or lowering the microscope objective.

METHOD'S VALUE: By using the visual confirmation method, the quantitative and qualitative determinations for virtual images become more exacting values. The immediate visual confirming observations eliminate all guesswork and prediction from the comprehension of the characteristics for virtual images.

STIMULATING INTEREST

IN REFLECTION

by Ronald I. Rank

Glenbrook South High School, Glenview, Illinois

PSSC Physics and other high school physics courses devote a proportionate share of coverage to physical optics. In any treatment of this topic, the reflection of light from plane or spherical surfaces is of major importance. The following describes a number of devices and techniques we find useful in exploring image formation of reflecting surfaces with our PSSC Physics students at Glenbrook South High School.

NOTE: These devices help stimulate and maintain student interest and thought about the principles of reflection. Several of them are featured in the classroom as year-round attention-getters.

Multiple Reflection

One subject which is emphasized in the PSSC course is the production of images as a result of multiple reflection from plane surfaces. As an aid in presenting this subject, mount two plane mirrors of fairly good quality in a corner of the classroom so that they meet at right angles to each other. The mirrors should be a minimum size of approximately 2½' x 3', and can usually be obtained from local furniture stores or Sears, Roebuck outlets.

When a student stands in front of this mirror arrangement, he sees three images of himself as shown in Fig. 1. Two are conventional plane mirror images

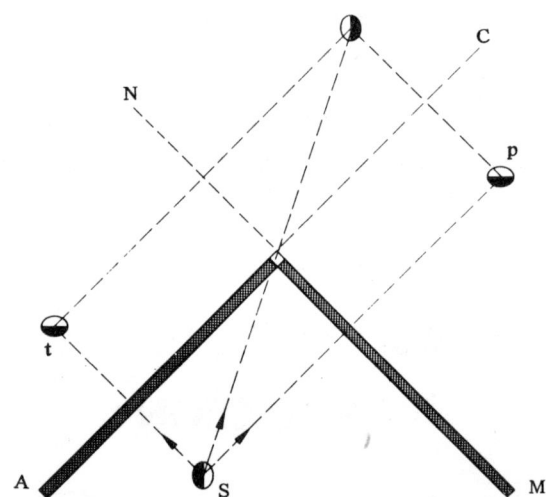

Fig. 1: *The reflecting surfaces are AC and MN. The image of object S in AC is t. The image of object S in MN is p. Image i is the image of image t to MN and also the image of image p in AC. These two images coincide when the angle between reflecting surfaces is 90°. Since image i is the result of a double reflection, it exhibits the characteristics mentioned in the text.*

but the third, seen at the junction of the two mirrors, is a bit unusual in that when the student raises his right hand, the right hand of the image is also raised. The image the viewer observes is truly that of himself in the way that others see him.

> NOTE: This device proves a good interest-rouser all year with our physics students. It is permanently mounted and is probably the first thing a student notices when entering the classroom.

Changing Angles

Some students will become interested in what happens when the angle between the mirrors is changed from 90°. They want to know how many images are formed and where they are located when the angle is changed to 45°, 60°, and so on. A device which is useful in discussing the effects of changing angles can be made available on the demonstration desk, and students may be allowed to explore with it any time they wish.

This device (Fig. 2) consists of a piece of wood 9" x 20" x¾" (dimensions

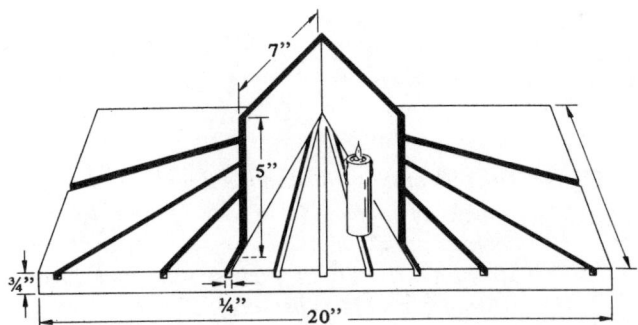

Fig. 2: *General view of the method of mounting the mirrors so that the angle between them may be varied by multiples of 15°. The saw cuts are all at angles of 15°.*

are not critical) with ¼" saw cuts (done in our wood shop with a radial arm saw) at angles of 15°. Two 5" x 7" x 3/16" front surface mirrors are mounted in two of the grooves so that they can be easily removed and reinserted into other grooves.

> SOURCE: The front surface mirrors are available from Edmund Scientific Company, Inc., 800 Edscorp Bldg., Barrington, New Jersey 08007 (Catalog 40043).

Since the angle between the mirrors can be changed by any multiple of 15°, the number of images and their character can be investigated individually by students. A candle serves as the object for the mirror combination. Another identical candle can then be used to locate the image by parallax. It helps to use crayon to color one-half of the object candle surface so that the effect of image reversal or nonreversal can be easily seen.

Sometimes students will attempt to see whether there is a relationship between the angle of the mirrors and the number of images formed. This can be investigated by making drawings to account for the images and their nature. For example, in Fig. 3 when the angle is 45°, the number of images is seven. The method used to locate all of the images is described in the caption for the figure.

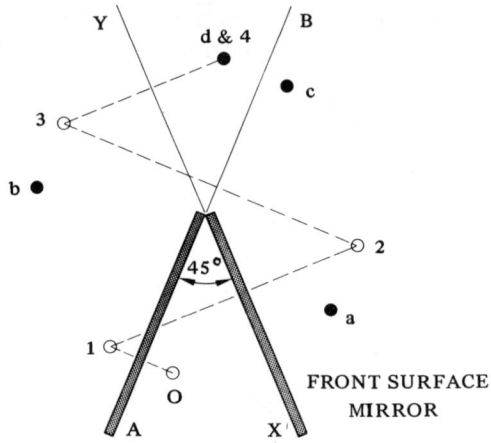

Fig. 3: *The reflecting surfaces are AB and XY. 1 is the image of object O in AB (found by the conventional method on a perpendicular from O to AB in such a way that image distance = object distance). 2 is the image of 1 in XY; 3 is the image of image 2 in AB, and 4 is the image of 3 in XY. Images a, b, c, and d are located in the same manner. Notice that image 4 and image d coincide.*

Parallel Reflecting Surfaces

A question usually arises concerning what happens when the angle between the two reflecting surfaces becomes 0° (mirrors are parallel). The number of images formed is, of course, infinite, depending upon the degree of precision involved in aligning mirrors. To demonstrate this phenomenon, a "pass-around" black box constructed as shown in Fig. 4 is helpful.

The black box consists of a box with a small flashlight bulb mounted in it between four mirrors, which are mounted on the inside surfaces. The mirrors should be front-surfaced to reduce the absorption loss to a minimum. Two small holes are cut in the walls of the box so that students can look into it from the

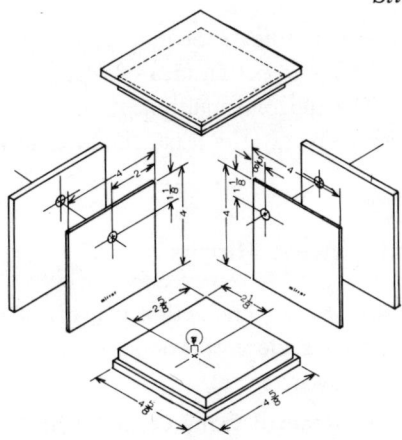

Fig. 4: *An exploded view of the box. The two front surfaces are not shown for reasons of clarity. Dimensions are in inches, and they will actually depend on the size of the mirrors used. Construction can be of plywood, Masonite combined with pine, etc., depending upon the builder's ingenuity. The bulb is mounted in a miniature porcelain socket (not shown) and the wires are passed through the bottom of the box to the switch (not shown).*

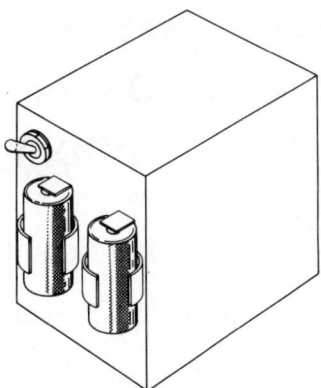

Fig. 5: *Overall outside view showing placement of the switch and batteries which are mounted on the outside.*

outside in such a way as to see the image of the lighted bulb reflected again and again and again.

HOLES: Holes for viewing through a mirror can be made in the reflecting surface by using one drop of dilute HNO_3 at the appropriate spot. The solution should be carefully blotted with a paper towel and then immediately rinsed with distilled water.

When a student looks into the box and lights the bulb by closing a switch mounted on the outside (Fig. 5), he begins to get an idea of what happens to a photon in a perfectly reflecting cube. The apparent multiplication of the number of sources is very interesting to observe. To say the least, the interior becomes very bright.

Convex and Concave Mirrors

A final interest-rousing device for study of reflection is made of spherical convex and concave mirrors mounted on a piece of wood with regular plastic mirror mounts (available in a dime or hardware store).

SOURCE: The mirrors may be obtained from Central Scientific Company, 2600 S. Kostner Ave., Chicago, Illinois 60623 (Catalog # 85417 - 16" diameter spherical convex mirror; #85407 - 16" diameter spherical concave mirror).

The mirrors are hung on a wall, preferably in the front of the room. Students will then see images in the mirrors from the first day they are in the classroom. When it is time to discuss the properties of the curved reflecting surfaces, small model airplanes can be suspended on strings from the ceiling in such a manner that the images will be seen by the students while they are seated.

AN INEXPENSIVE APPARATUS FOR DEMONSTRATING COLOR ADDITION

by Charles C. Smith

Malvern Preparatory School, Malvern, Pennsylvania

Ask the average student to name the three primary colors and also list the colors produced when these primary colors are added in various combinations. Very likely he will give a satisfactory answer for the subtractive effects with red, yellow, and blue as primaries. But he probably will not even be aware that there is a difference between the additive and subtractive processes.

And if he does know of this difference from reading about it in his text, he probably doubts that it really exists. After all, anyone knows you can't get white paint by mixing red, green, and blue paint in a can! Most of the student's experience has involved the subtractive process; the additive process will not be real to him until he sees a simple, convincing demonstration. Even then he will be skeptical and retain some mental reservations! Though he acknowledges that the color TV tube is real, and it adds red, green, and blue to get white, there is too much of the "black box" concept in it to convince him.

NOTE: A variety of apparatus for demonstrating the additive process is available from supply houses, but they are expensive and use either three independent light sources or a complicated lens and mirror system with adjustment problems. However, the homemade apparatus described here is easily constructed and inexpensive if you have on hand a 2" x 2" slide projector and a small 300-RPM, 110-volt A.C. motor.

Working principle: The principle behind the apparatus is quite simple. A disc with red, green, and blue filters is rotated at high speed in the beam from the projector. The succession of red, green, and blue sensations is so rapid that the eye and brain are only aware of the additive white produced so long as the object illuminated is not in motion. For the student the convincing aspect is that he can see clearly that only red, green, and blue light are used, yet also that the result is white and not the three colors.

The white light produced by the apparatus will illuminate a fairly large area with sufficient intensity to present a variety of exciting demonstrations in your classroom or in the auditorium. Some possible demonstrations are suggested later in this article.

Constructing the Addition Apparatus

Securing the light source: It is most convenient to make a single unit with the motor mounted firmly in place to provide support for the slide projector, and also an electric outlet for both the motor and the projector. Fig. 1 shows the design and dimensions of the unit we constructed. To accommodate your projector, you may find it necessary to modify the dimensions given.

(1) For building the support, you can use scraps of 1-inch white pine and ¼-inch plywood. Note that the vertical distance from the center of the projector lens to the motor shaft must be about 1¾ inches less than the radius of the filter disc. Keep this in mind when planning the height of the motor support.

(2) Determine the dimensions of the base to which the motor support is firmly bolted by the basal dimensions of your projector. This base, too, can be of 1-inch white pine.

(3) Mount a double outlet with a long extension cord on the side of the motor support. Both the projector and motor cords are plugged into this outlet when the apparatus is in use. An "off" toggle switch may be mounted near the motor so that it can be separately controlled.

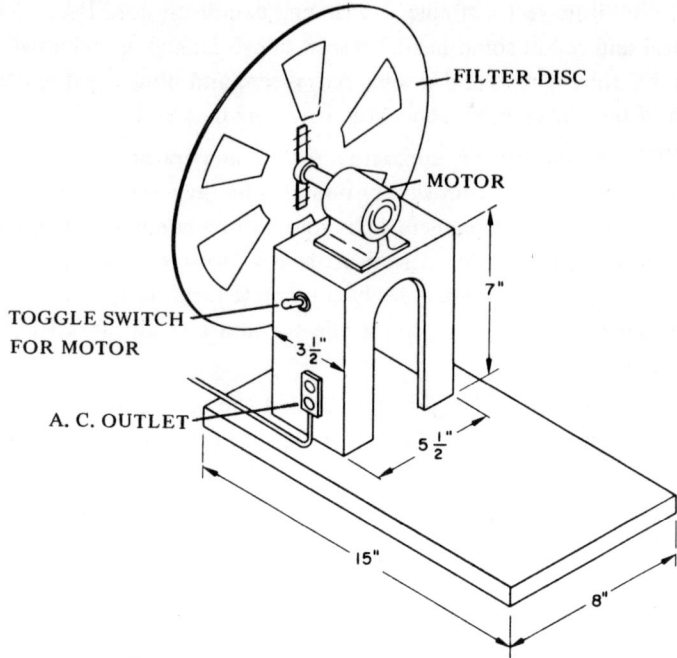

Fig. 1: Motor and Projector Unit

Making the filter disc: Since the disc will rotate at 300 RPM, it must be quite stiff and have no tapes or filter edges to tear loose in the resulting air friction. Standard poster board is a satisfactory material for the disc.

(1)   Cut out two duplicate patterns from the poster board as illustrated in Fig. 2, determining the dimensions of the filter apertures by the size of the projector lens.

(2)   After cutting the filters to a size slightly larger than the filter apertures, Scotch-tape them in place and then cement the two discs together, properly registered, and with the filters between the two discs.

(3)   To prevent warping, keep the discs flat, under pressure, until they are thoroughly dry.

COLOR FILTERS: The red, green, and blue color filters can be cut from colored cellophane; however, a more convincing white can be gotten from inexpensive filters purchased from Edmund Scientific Corp., Barrington, New Jersey 08007. (You may want to experiment with a variety of colors on a makeshift disc before making a permanent one.)

Mounting the disc: Mounting the disc firmly on the motor shaft poses a problem. Since we wanted to use a set of interchangeable discs during a demonstration, we had to come up with a means for making a simple, rapid

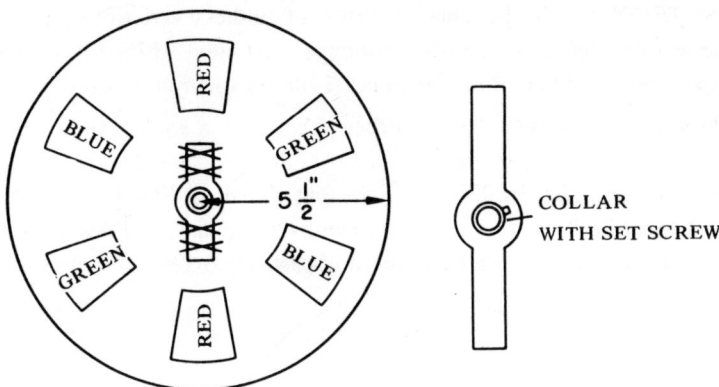

Fig. 2: Filter Disc   **Fig. 3: Filter Disc Support**

exchange. After some experimentation, we designed a collar and support as shown in Fig. 3. In this arrangement the set screw holds the disc firmly on the shaft and the plate support holds the disc to the collar without tearing off at high speed.

If you use this mounting device, make certain that the collar is exactly centered on the disc, then tape it in place and reinforce the tape with thin wire lacing.

VIBRATION: If there is excessive vibration when the motor is driving the disc, it is not properly centered and you may have to do some careful trimming.

A cooperative industrial arts teacher in your school will be glad to make several of these supports for you out of brass scraps, or you may be ingenious enough to improvise your own method of attaching the disc to the motor.

Testing the motor: You should now be ready to put the projector in place and test its operation. The projector lens should extend into the motor support so that it is within an inch of the rotating disc. If you cannot prevent the entire apparatus from "creeping" while the motor is running, clamp it to the table with a C-clamp.

Demonstration Suggestions

Procedure 1: Hold your hand in the beam of white light. It will appear quite normal. Then flip your hand as if you wanted to shake off water, and your students will see a jumbled mass of red, green, and blue fingers sharply outlined. Hold the hand still and the colors will disappear. Some students will note that if they blink their eyes rapidly, they will still get fleeting glimpses of color.

As a variation, show students a dowel stick about a foot long which has been painted white. Swing the stick rapidly back and forth through a 90-degree arc and it becomes a three-colored fan. Or toss a white ping pong ball across the beam; it becomes a chain of red, green, and blue balls.

NOTE: You will be able to think of endless variations on this procedure by applying the principle that any white object will appear white when not in motion, tricolored when in motion.

Procedure 2: If you have another small motor with a variable speed or perhaps a hand rotator, you can prepare even more spectacular effects. Cut out a large disc, about 2 feet in diameter, like that shown in Fig. 4. (A large sheet of black cardboard will serve very well.) Paint on the diamond design with white paint, or cut out white paper diamonds and paste them onto the cardboard.

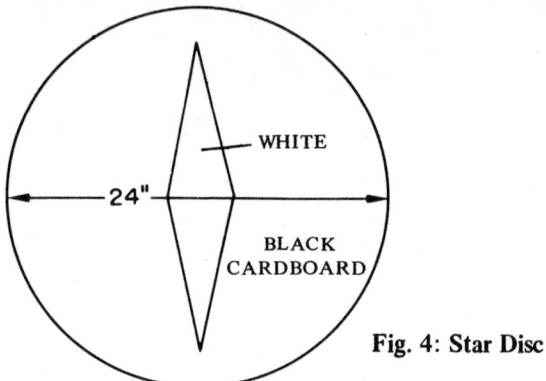

Fig. 4: Star Disc

If you use the disc with a hand rotator, cut out a hole in the center for mounting the disc on the rotator shaft. If you use a small motor with a rheostat control, you will need a collar device similar to the one holding the filter disc. The support for the motor should be tall enough so that the disc will clear the table and should have a base so that it can be clamped firmly to the table.

When this large disc is completed, place it so that the white light from the projector illuminates the entire disc. Then, rotate the disc at various speeds to produce a variety of color effects. At one speed, due to the stroboscopic effect, the students will observe a many-pointed star apparently at rest.

OBSERVATIONS: The points of this star will be a succession of red, green, and blue colors. Its center will be a white circle. The overlap of the red and green points will produce a small yellow triangle; the green and blue, a small cyan-blue triangle; and the blue and red, a small magenta triangle. (This is particularly convincing to students since they can see directly the effect of adding three primaries, or of adding them in pairs.)

By varying the relative speed of the two discs, you can produce these colored stars with varying numbers of points. If you make a second filter disc using alternating yellow and blue filters, you can repeat the preceding demonstrations to show students that adding yellow to blue does not produce green, but white.

Procedure 3: To pursue the variety of demonstrations further, make another black disc for the rotator. Cut out four 2-inch white circles and paste them onto the disc 90 degrees apart about 2 inches from the margin. This will produce a circle of overlapping circles, also demonstrating the additive effect. A 5-inch square of white paper pasted in the center of the disc will add to the effect.

Comparing effects: To contrast the additive effect with the subtractive effect, prepare several 2" x 2" slides by overlapping various combinations of the same filter materials and placing them in suitable mounts. Using the same projector with a screen, you can reassure students that they were correct in their original answer for one kind of color combination—if the process involved successive subtraction from a white light source—but that they must expect quite different results when they combine a number of different-colored sources.

COLOR EXPERIMENTS
WITH INEXPENSIVE EQUIPMENT

by Herman H. Kirkpatrick

Roosevelt High School, Des Moines, Iowa

Beautiful color displays are a fascinating way to introduce physics students to a study of the phenomena of light. I find that with the use of ordinary 35 mm slide projectors and inexpensive supplementary materials, I can easily demonstrate such topics as *color addition, color subtraction, reflection, refraction, dispersion,* and *diffraction.*

IMPORTANT: The experiments described here can be used as an introduction to the study of "Waves and Light" in the PSSC Physics course, or are readily adapted to the study of light as presented in any introductory physics course.

In addition to the slide projectors, here are the materials needed for the experiments described in this article:

Roscolene plastic filters (swatch book, obtainable from Rosco Laboratories, Inc., 214 Harrison Ave., Harrison, N.Y. 10528. Current price, 50¢).
Kodak ready mounts, 35 mm.
Diffraction grating, obtainable from Edmund Scientific Co.
Diffraction screen (200 mesh).
Flat black lacquer and 2" x 2" glass slides.
Small (compact size) mirror.

Experiments

1. *Color Addition:* Cut the Roscolene plastic filters to fit in the Kodak ready mounts and apply a warm iron to the edges of each ready mount. The slides thus prepared are ready for the projector. The plastic filters provide 36 colors for a wide variety of color addition experiments. For example:

(a) Place medium red filter #9/16 in one projector and medium green filter #9/40 in another. Aim the two projectors at the screen so that they overlap slightly. Yellow light will be observed in the overlap area. Similarly, magenta in the overlapping areas will be produced by using medium red #9/16 and medium blue #9/135. And blue-green or turquoise will be produced by using medium green #9/40 and medium blue #9/135.

(b) You can produce white light by using three projectors and over-lapping the three filters used in the first experiments described.

(c) You can produce white light by using two projectors to add light beams filtered through complementary colors.

2. *Color Subtraction:* Here is one of a number of interesting experiments that can be performed.

(a) Fold an ordinary index card and cut to an appropriate size to fit a 35 mm slide, 35 mm x 38 mm (or 1¼" x 1½"). Using an ordinary notebook punch, punch seven holes in the folded cardboard as shown in Fig. 1.

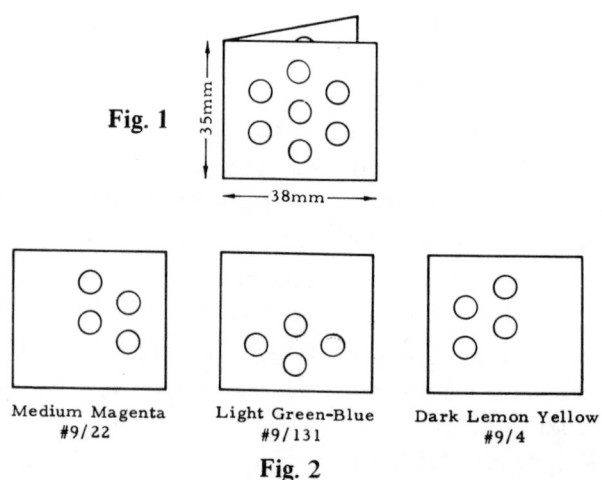

Fig. 1

35mm

38mm

Medium Magenta
#9/22

Light Green-Blue
#9/131

Dark Lemon Yellow
#9/4

Fig. 2

(b) Using the cardboard as a sort of stencil, select three filters (medium magenta #9/22, light green-blue #9/131, and dark lemon yellow #9/4) and punch only selected holes, as shown in Fig. 2. The cardboard will permit you to have perfect alignment of the punches in all three filters.

(c) Place all three filters one on top of the other in a Kodak ready mount and show on the screen. Colors will show as illustrated in Fig. 3. Note that holes have been punched in the center of all three filters, so that unfiltered (white) light shows in the center. Then, starting with the yellow light in the lower right-hand corner of Fig. 3, here is how the combinations work:

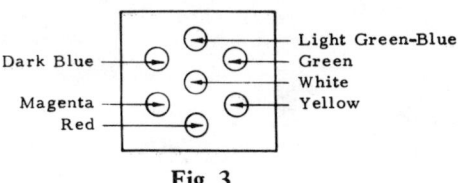

Fig. 3

The light reaching the yellow circle on the screen has passed through a yellow filter only.

The light reaching the green circle has passed through both a light green-blue (turquoise) and yellow filter.

The light reaching the turquoise circle has passed through a turquoise filter only.

The light reaching the dark blue circle has passed through both a turquoise and magenta filter.

The light reaching the magenta circle has passed through a magenta filter only.

The light reaching the red circle has passed through both a magenta and yellow filter.

NOTE: The 36 colors offered by the Roscolene plastic filters give you a wide variety of subtraction experiments in addition to the one described.

3. *Reflection:* Paint a 2" x 2" glass slide with a quick-drying black lacquer. When dry, make a fine line in the paint with a razor blade. (You can prepare slides with varying widths of slits by making two marks and carefully scraping the paint between the lines with a razor blade.) Place the slide in a projector with the slit in vertical position. By placing a small mirror in front of the projector you can reflect the image of the slit on the screen and, if you wish, measure the incident and reflection angles.

4. *Refraction and Dispersion:* With one of the previously prepared 2" x 2" glass slides, you can place a prism in front of the projector and observe refraction and dispersion. Maximum resolution of the projected spectrum may be obtained by carefully adjusting the prism for minimum deviation and by focusing the lens. (Fig. 4.)

5. *Diffraction:* By using 2" x 2" slides with a slit, or slits, prepared as

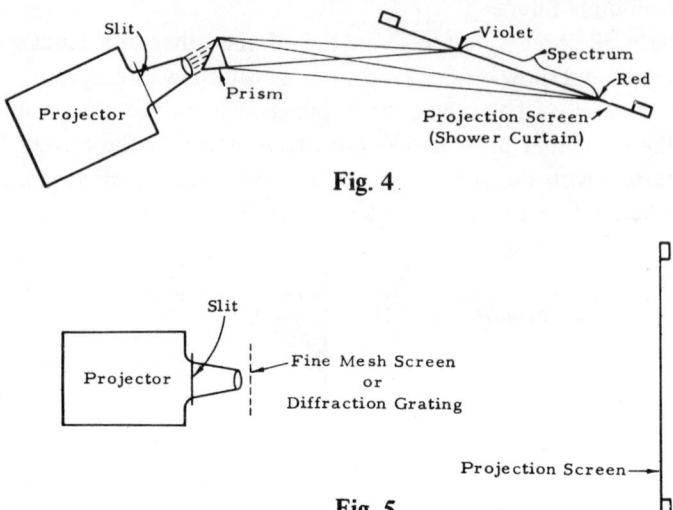

Fig. 4

Fig. 5

explained, you can obtain interesting diffraction effects by placing slits, double slits, screens, and diffraction gratings in front of the projector lens (Fig. 5).

NOTE: Diffraction experiments will probably require a carefully darkened room if done as classroom demonstrations.

Quantitative data may be readily collected for you to calculate the wave length of the various colors of light.

4

WAVES

INEXPENSIVE RESONANCE TUBE SETUPS FOR SOUND LABS

by Harvey H. Radke

Washington Park High School, Racine, Wisconsin

Following is a design for inexpensive resonance tube setups that we have used with success in our sound labs. Each apparatus can be easily put together at a cost of about $1.00, making use of some laboratory hardware. The design is illustrated in a general drawing (Fig. 1) and two specific area drawings (Fig. 2).

Overall Design

A ring stand is used for support of the system, shown in Fig. 1. R_1 and R_2 represent points at which a ring or utility clamp could be placed for tube support. However, if a ring is used at R_2, a can cover or coffee lid with a hole in it the size of the Tygon tubing should be placed on the ring first to support the tube. Another ring may be located at point R_3, if desired, as a storage place for the reserve water supply.

Specific Fittings

The original top of the plastic bottle is fitted as shown in Fig. 2, D_1. Directions for this assembly are as follows:

1. Place a hole in the original bottom of the bottle.

    NOTE: An easy way to do this is by heating a piece of metal, such as iron pipe, in a Fisher burner until the iron is quite hot, and then pushing the hot metal against the bottle's bottom.

2. Put the Tygon tubing through the stopper as shown in the drawing, then drop it into the bottle through the new hole and pull it out the other end until snug. (Glass tubing can be used instead of Tygon, but it makes the setup breakable.)

    To insure that the stopper assembly will drop properly, insert a wire into the original hole and through the new hole, slide the tubing/ stopper assembly onto the wire, and drop it into its final place.

92

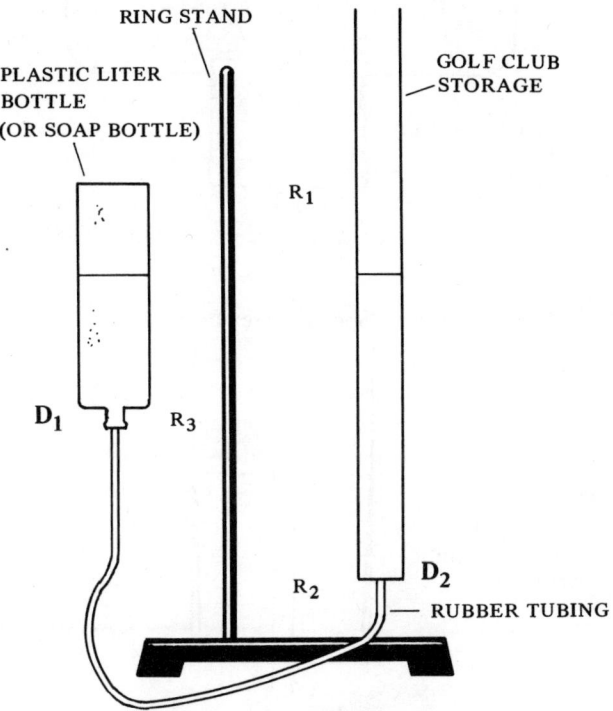

Fig. 1: General Drawing

3. Seal the assembly with sealing wax, as shown in Fig. 2, D_1.

NOTE: Do not overheat the sealing wax or it may distort the Tygon and the bottle.

4. Put together the second assembly following the illustration in Fig. 2, D_2, and the preceding general directions.

Usage

Many standard physics and physical sciences texts discuss this type of equipment and provide related experiments. For example, such material can be found in *Modern Physics* and the accompanying lab manual, and in the paperback publications, *Visualized Physics* and *Graphic Survey of Physics:*[1]

[1]John E. Williams, *et al, Modern Physics.* (New York: Holt, Rinehart & Winston, 1968), pp. 296-7. Manual: *Laboratory Experiments in Physics*, pp. 56-7.

Alexander Taffel, *Visualized Physics*, rev. ed. (New York: Oxford Book Company, Inc., 1954), pp. 161-2.

Alexander Taffel, *Graphic Survey of Physics*, rev. ed. (New York: Oxford Book Company, Inc., 1963), p. 185.

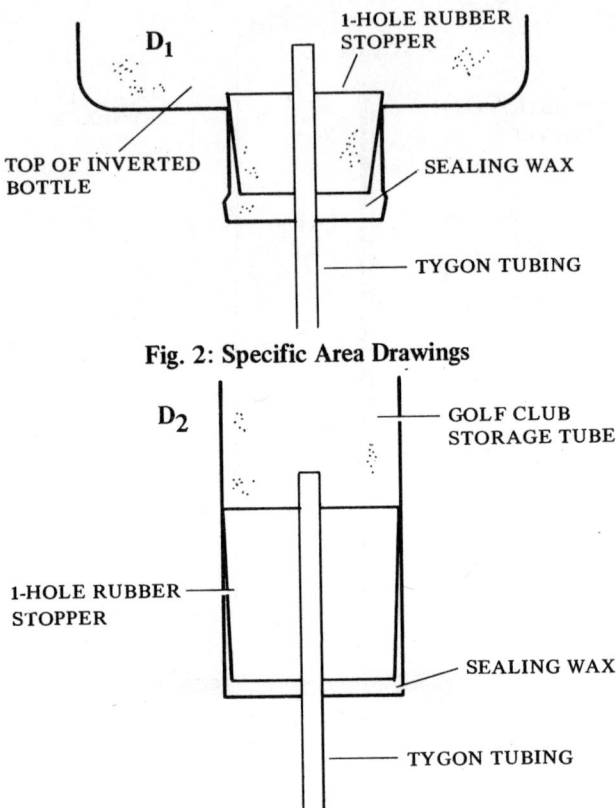

Fig. 2: Specific Area Drawings

Following is an example of my use of this equipment in a noncollege preparatory physical science lab. (The material is too simplified for use with more advanced and/or high ability students who are experienced in lab work; among other things, the standard formula letters have been changed to English, and there is no correction made for resonance tube diameter.)

Experiment: The Length of Sound Waves

Introduction: Sound waves may be bounced off objects much as a ball is bounced off a wall. The reflected, or bounced, sound wave is called an echo. If such an echo joins another sound wave directly from a sound source, we say that sound has been re-enforced, or that there is resonance. This would make either of the two sounds sound much louder. In this experiment we will use the idea of resonance to measure the length of sound waves.

Apparatus: Arrange your equipment as shown in Diagram 1.

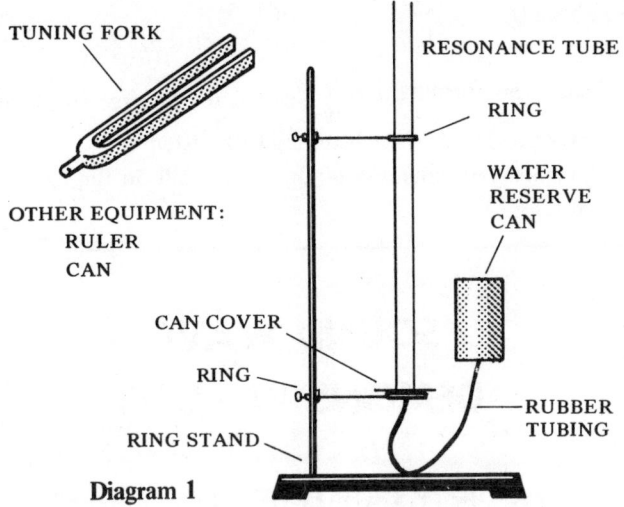

Diagram 1

Procedure:

(1) Raise the water reserve can above the top of the resonance tube. Then, using another can, fill the resonance tube full of water. Now by lowering or raising the water reserve can, you will see that because water always seeks its own level you can easily regulate the water level in the resonance tube.

NOTE: The tuning fork is constructed so that when it has been started vibrating it will vibrate at a uniform speed, and thus produce sound waves of one wave length.

(2) With the resonance tube full of water, strike the tuning fork on a rubber stopper and move it to the top of the resonance tube. Next, slowly allow the water to lower in the resonance tube. Every few seconds, as the fork's amplitude of vibration becomes small, stop the water movement down the tube and restrike the fork. Then continue on downward.

(3) As the water level changes, suddenly you will hear the sound become louder for a moment. Resonance has occurred. Raise the water level a little and lower it carefully to locate the spot precisely. *Remember to keep the tuning fork vibrating.*

(4) Measure the distance from the water level to the tuning fork. This is the length the sound waves resonated in, ¼ the length of sound waves produced by the tuning fork.

Calculations:

a. Calculate the length of each tuning fork wave tested.

b.   Calculate the speed of sound at room temperature:

$$1090 \text{ ft/sec at } 0°C$$

(2 ft/sec for each
degree above 0)

+ _____

_____ ft/sec at ____ °C

c.   Using the formula $F = \dfrac{V}{W}$, calculate the frequency of the sound waves produced by each tuning fork tested.

d.   Compare your answers to those stamped on the base of the tuning forks.

SOUND WAVE DEMONSTRATIONS

by Rudolph E. Petrucci

Cranston High School West, Cranston, Rhode Island

Here are three demonstrations whose general purpose is to present to students visually what they perceive audibly. Specifically, the objectives of the demonstrations are:

(a)   To exhibit the waveforms produced by tuning forks on the screen of a cathode ray oscilloscope (CRO).

(b)   To demonstrate what is meant by the amplitude (or intensity), frequency, and wave length of a sound wave.

(c)   To demonstrate the effects of mixing or "beating" two frequencies together.

Apparatus

The following apparatus will be needed to perform the three demonstrations:

2 tuning forks (256 vps and 512 vps)
1 rubber mallet
1 cathode ray oscilloscope
1 audio amplifier with speaker and microphone

NOTE: A 16-mm movie projector (RCA or Bell & Howell) with a microphone will do nicely.

2 audio signal generators

Exhibiting Waveforms

Use the following procedure to demonstrate waveforms:

Step 1: Set up the apparatus according to Fig. 1.

NOTE: Be certain to orient the speaker away from the microphone in order to prevent regenerative feedback which will cause loud squeals or howls. (A real attention-getter is produced by moving the microphone in front of the loudspeaker, then moving it away and towards the speaker. This creates a series of oscillations in a regular pattern.)

Step 2: Adjust the loudness (volume) control on the amplifier to amplify the vibrations from the tuning fork.

NOTE: Do not advance to the limit, as excessive noise may occur.

Step 3: Adjust the vertical amplifier gain on the CRO for a "2–3" pattern.

Step 4: Strike the 512 vps tuning fork with the rubber mallet and hold it near the microphone.

NOTE: It is best to have a tuning fork which is mounted on a resonator. If not, one can secure the tuning fork by means of a right-angle clamp to a supporting stand.

Step 5: Choose the correct coarse frequency (sweep) range and adjust the fine frequency control on the CRO in order to obtain two stationary sine waves. The tuning fork will have to be re-energized several times.

Step 6: Repeat Step 5 until a good pattern is obtained. The synchronizing mode should be on internal and the synchronizing control set to about half its range.

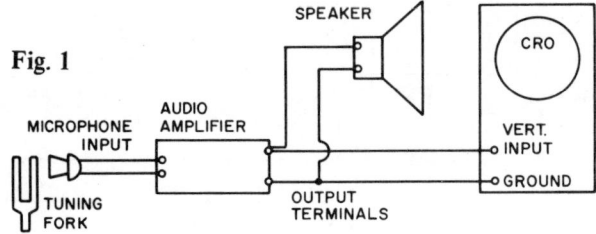

Fig. 1

Explanation:

Since the period or time for one cycle is t (t equals $\frac{1}{f}$), the distance on the cope face from one peak to the other represents time. In this application, $= \frac{1}{512}$ seconds or .00195 sec. As the tuning fork's energy decays, the amplitude or height of the waveform does likewise, but the number of cycles remains constant.

Demonstrating Amplitude, Frequency, and Wave Length

Step 1: Without making any adjustments on the CRO, substitute a 256 vps tuning fork and again energize it with the rubber mallet. Note that the tuning fork may have to be moved closer to the microphone.

Step 2: If the scope is properly adjusted, only half as many waves will appear. If this does not occur, a slight adjustment of the fine frequency control should be made.

Explanation:

The 256 vps tuning fork vibrates at half the frequency of the 512 vps tuning fork and, therefore, will produce only half as many sine waves. Striking the tuning fork harder will cause a waveform of greater amplitude to appear but it will not produce more sine waves.

Demonstrating Beat Frequencies

Step 1: Disconnect the amplifier and the microphone from the circuit and replace these with the two signal generators as shown in Fig. 2. Note that they are in parallel.

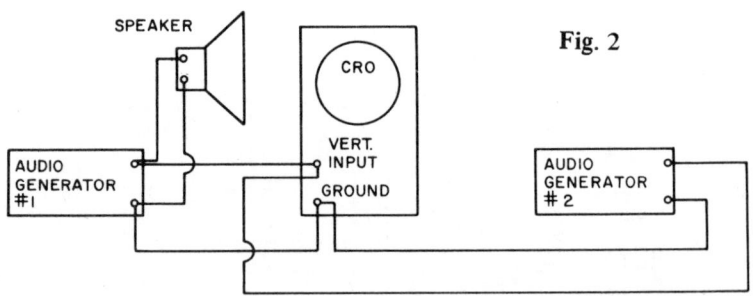

Step 2: Set one signal generator to 300 cps and increase the output level so that it can be easily heard in the speaker and seen in the CRO.

NOTE: Do not increase to the point of distortion.

Step 3: Turn off generator #1 and repeat step #2 with generator #2.

Step 4: Turn on generator #1, and with number 1 generator's frequency fixed at 300 cps, carefully adjust the frequency of generator number 2 so that both audio generators are in step.

NOTE: The setting may be slightly off. Indicate to the students that this is one way to calibrate an instrument to a standard.

There will be a slight wobbling due to the fact that each generator is free running.

Step 5: Again adjust the fine frequency control on the scope to obtain two to four stationary sine waves.

Step 6: Now slightly increase the frequency of generator number 2 so that a distinct beat frequency is heard. Observe the waveform.

NOTE: A slight adjustment of the synchronizing control and/or the fine frequency control may have to be made in order to obtain the waveform as shown in Fig. 3.

Step 7: Again increase the frequency of generator 2 so that two distinct tones are heard on the loudspeaker. The waveform should indicate constructive and destructive interference of two waves similar to those in Fig. 4.

Fig. 3

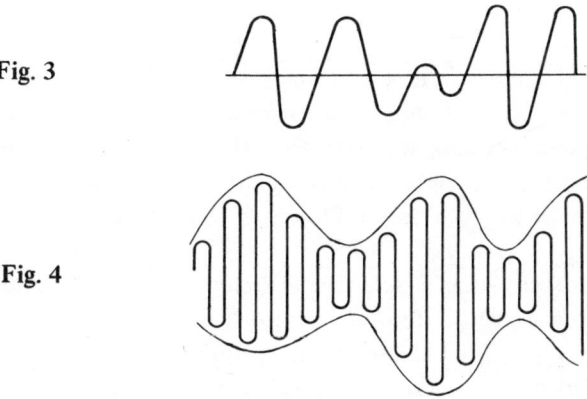

Fig. 4

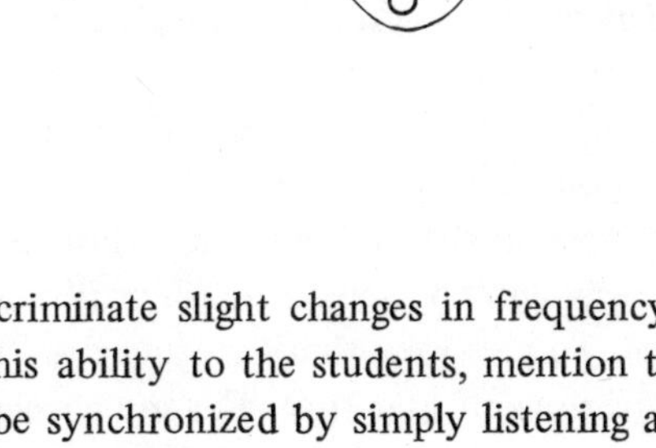

Explanation:

The human ear can discriminate slight changes in frequency very effectively. To further illustrate this ability to the students, mention that the twin engines in a power boat can be synchronized by simply listening and adjusting the speed to eliminate the beat frequency.

Whenever two different frequencies occur simultaneously, some parts of the waves are in phase and therefore are additive. On the other hand, other parts are out of phase and consequently are diminished. (If generator 2 is set at 2X, 3X, 4X, etc., the frequency of generator 1, harmonics can also be displayed.)

Making Comparison Checks

If an amateur radio receiver capable of receiving 2.5, 5, 10, 15, 20, or 25 mega-hertz is available, frequency comparison checks can be made with one of the audio generators by means of Lissajous figures. The National Bureau of Standards operates Station WWV on the above frequencies. In addition to time signals, standard audio frequencies of 600 cps and 440 cps are also transmitted. These are accurate within 1 part in 10^8.

SIMULATED WAVE MOTION
AND THE DOPPLER EFFECT

by William Naison

Formerly Jamaica High School, Jamaica, New York

It is good physics to have students base definitions and derivations upon operations performed by them. I originally sought a teaching device in order to derive relations concerning the Doppler Effect. The device described in this article turned out to be useful in teaching operational concepts of wave motion without reference to the Doppler Effect, as will be seen in the first example given.

All that is needed is a steel tape and clothespins, and a student to act as timekeeper. The initial position of the tape is marked on a blackboard, and as the steel tape is pulled out at a regular rate, clothespins are attached to the tape. At some designated time and place along the path of the tape, a student picks off the clothespins while another student holds the end of the tape with sufficient tension to maintain a horizontal position.

The method of timing is designed to slow up the action, or to stop it entirely, to permit discussion and resumption of the demonstration after the discussion. This is done by selecting, arbitrarily, unit time to be any convenient length of time. To avoid fractional frequencies, the timekeeper is asked to call out suitable fractions of unit time. For my purposes, unit time was divided into quarters, and the timekeeper was asked to call out, in succession, one-quarter unit time, two-quarters unit time, and so on.

The operational definition of frequency is obtained by having the student count the number of clothespins picked off in unit time. It is possible for the whole class to perform this operation by having each student count the number of pins passing by a visible marker. The actual measurement of the distance between pins adequately defines the wave length and the time interval between the arrival of successive pins defines the period. The reciprocal relation between frequency and period is empirically obtained. The velocity of the wave may be calculated by substituting wave length for distance, and period for time, in the ordinary definition of velocity (distance divided by time).

NOTE: **Prior to the demonstration, you must decide exactly what the clothespins are to represent. Initially, each pin represents maximum displacement of the medium (crest of a wave) which travels through the medium at a finite speed. Ultimately, the meaning can be extended. The pin might stand for any given phase.**

Example 1—Stationary Source and Stationary Observer

Basic Facts Provided: The speed of the wave in the medium is 20 inches per unit time. The unit of time is 40 seconds. The timekeeper calls out the time in quarter units of time every ten seconds. Students were told that the source was producing a periodic disturbance but the actual frequency of four cycles was not revealed. (For convenience in handling, the first 5 inches of tape was drawn out of the tape reservoir.)

Procedure: At the instant the timing is started, the teacher (or student) attaches a pin at the 5-inch mark. The tape is pulled out 5 inches during the first quarter of unit time (based upon the given velocity of 20 inches per unit of time) and again a pin is attached to the tape; this pin is at the 10-inch mark. The procedure is repeated until four pins have been attached.

At a reasonable distance from the source—say 20 inches—an observer is stationed to count the number of pins passing by in unit time. The source continues to attach pins until the observer's count is complete; that is, until unit time has passed for the observer.

Example 2—Motion of a Source Toward a Stationary Observer

For reference purposes, the source is held stationary during the first two-quarters of unit time, following the same procedure and data as in Example 1 above. The action is then stopped to announce that the source will move with a velocity of 10 inches per unit time (or 2½ inches per quarter of unit time). During the third quarter of unit time, when action is resumed, the second pin moves 5 inches from the original position of the source. Up to this time the source has been stationary, but now it begins to move until it has traveled a distance of 2½ inches in the same direction as the pins. The new position of the source is marked on the blackboard and a new pin is attached—it will be only 2½ inches from the pin ahead.

NOTE: Although the motions of the pins and the source should have taken place simultaneously, they are performed sequentially. Justification for this can probably be obtained from your students who have a knowledge of vector addition.

The action of the moving source is continued until a suitable stationed observer has completed the count of pins passing by in unit time.

To derive a formula for the observed frequency, begin by using

$$F_o = \frac{V_o}{L_o}, \qquad \text{(Formula 1)}$$

where the subscripts refer to the observed values of frequency, velocity, and wave length with a moving source. Confusion may result unless the observed velocity is defined as the relative velocity between the observer and the wave crest (the pin).

The students will note that it is the motion of the source that has brought about the observed change in wave length. If the source were stationary, the distance between the crests (the wave length) is equal to the velocity of the wave crest, relative to the ground, multiplied by the period (distance is the product of velocity and time). When, however, the source is moving relative to the wave crest, the *velocity of the wave crest relative to the source* must be multiplied by the period.

The reduced wave length which the observer measures can be expressed as:

$$L_o = P\,(V\text{-}V_s) \qquad \text{(Formula 2)}$$

L_o is the observed wave length
V is the velocity of the wave crest relative to the ground or a stationary source
P is the period (the reciprocal of the constant frequency of the source)
V_s is the velocity of the source relative to the ground.
The observed values in Formula 1 can now be replaced by their equivalents, as follows:

$$F_o = \frac{V}{P\,(V\text{-}V_s)}. \qquad \text{(Formula 3)}$$

The student confirms by observation that the observed velocity of the wave crests is identical to that observed with a stationary source. Consequently, the numerator in Formula 3 may be replaced by the velocity of the wave crests relative to the ground. And, in addition, if the period is replaced by the reciprocal of the frequency of the source, Formula 3 then becomes:

$$F_o = P\,\frac{V}{V\text{-}V_s}. \qquad \text{(Formula 4)}$$

Example 3—Observer Moving Toward a Stationary Source

The procedure at the source and all the data, except for the velocity of the observer, is the same as in Example 1. The procedure at the observer's location must now be modified to take into account the observer's motion, which is chosen to be 20 inches per unit time.

The observer is stationed next to a wave crest (pin) when the count is ready to begin. During the first quarter of unit time, this crest moves 5 inches away from the observer and another wave crest moves into alignment with the observer. The observer remains stationary during this action. The wave crests now remain stationary while the observer moves a distance of 5 inches toward the source; this action brings the observer into alignment with still another wave crest.

NOTE: As in Example 2, the motions of the wave crests and the observer are simultaneous but are demonstrated sequentially within the same time interval.

Each new position of the observer is marked and the procedure is repeated until unit time has elapsed.

The derivation of a formula to calculate the observed frequency begins, as in Example 2, with Formula 1, which is now applied to a moving observer. Investigation by the students reveals that the observed wave length is no different from that observed by a stationary observer, as in Example 1. The observed velocity, however, is different. The students note that the observed velocity is the sum of the velocity of the observer relative to the ground (or the source) and the velocity of the wave crest relative to the ground (or the source). Thus, Formula 1 may now be expressed as:

$$F_o = \frac{V + U}{L_o} \qquad \text{(Formula 5)}$$

V is the velocity of the wave crest relative to the ground
U is the velocity of the observer relative to the ground.

Since the observed wave length has not changed from that seen under stationary conditions, as in Example 1, the observed wave length may be written as:

$$L_o = \frac{V}{F} \qquad \text{(Formula 6)}$$

F is the source frequency
V is the velocity of the wave crest relative to the ground.

Substitution of Formula 6 in Formula 5 yields:

$$F_o = \frac{F(V + U)}{V} . \qquad \text{(Formula 7)}$$

STANDING WAVES
ON A VIBRATING STRING—
DEMONSTRATIONS

by Melville R. Bosse

Withrow High School, Cincinnati, Ohio

In the second edition of *University Physics** there is a description of a method by which standing wave patterns on a vibrating nonmagnetic conductor can be amplified and transferred into an oscilloscope picture. Following are some interesting demonstrations based upon this method.

The Motor Principle

According to a discovery made by Hans Christian Oersted in 1819, a force will exist between a current-carrying conductor and a magnetic field—the so-called motor principle which is also applicable in the operation of electro-dynamic loudspeakers and many other devices. This force can be demonstrated as follows:

1. Stretch a long, *stranded* copper wire free to vibrate between the poles of a strong permanent magnet as shown in Fig. 1. (This experiment was originally done using solid iron, brass, and copper wires and it was found, from trial and error experience, that stranded copper works best. Iron is magnetic. Solid copper and brass break too easily.)

NOTE: #18 or #20 bell under 50 to 100 Newtons of tension will do nicely. The wire may be purchased as Stranded Bell Wire from Macalaster Scientific Corp., 186 Third Ave., Walton, Mass. 02154.

2. Pass a current of 5 to 10 amperes D.C. through the wire. A satisfactory power supply is a Lab Volt, Model 190 (obtainable from the Buck Engineering Company, 37 Marcy Street, Freehold, New Jersey 07728). It is continuously variable from 0 to 10 volts, 0 to 10 amperes D. C.; or from 0 to 20 volts, 0 to 5 amperes A. C. The exact setting will depend upon the length and resistance of the wire,

---

*F.W. Sears and M.W. Zemansky, *University Physics,* 2nd ed. (Reading, Mass.: Addison-Wesley Publishers Co., Inc., 1957), p. 374.

Fig. 1: An easy way of obtaining and adjusting from 5 to 100 Nt of tension.

the strength of the large permanent horseshoe magnet, the amount of tension used, and the desired amplitude—which will be large enough to be seen but not violent enough to break the wire.

The current will visibly force the wire to move in a plane perpendicular to the direction of the magnetic field. Reversing the direction of the current (or the field) will reverse the direction of the motion. (With such a showing, the teacher may introduce the right-hand rules for magnetic deflection.)

If a current of 5 to 10 amperes, 60 Hertz A.C. is passed through the wire, it will vibrate at the same frequency. Adjustment of either tension or length to a resonant condition will greatly amplify the vibration (Figs. 2 and 3).

NOTE: Care must be taken when using a weaker solid wire conductor, for the wire may snap under continual strain and bending.

Determining Wave Velocity

A careful measurement of wave length (twice the distance between the nodes in the standing wave pattern) multiplied by the frequency will give an indication of wave velocity. This can be verified by the equation $v = \sqrt{T/u}$, where v is the velocity of a transverse wave in a vibrating string, T is the tension in the string in Newtons, and u is the mass per unit length of the string in kg/m. The last is easily determined with a meter stick and a triple beam pan balance.

With the magnet and current removed, the wire can be plucked and a sound of the same frequency can be heard. The frequency of the sound heard

Fig. 2: A wire carrying 5 amperes at 60 Hertz A.C. will vibrate with considerable amplitude if stretched between the poles of a permanent magnet and if the natural frequency of the string is the same as, or is a harmonic of the frequency of, the alternating current (resonance).

WOOD STOP TO ADJUST VIBRATING LENGTH.

Fig. 3: The same wire when adjusted under the same conditions to a nonresonating frequency will show no visible signs of vibration.

will be the same as the frequency of the vibrating string (wire). However, since the speed of sound in air is *not* the same as the speed of a transverse wave in the vibrating wire, the wave length of the sound in air will not be the same as the wave length (twice the distance between stops when vibrating as a whole) of the corresponding vibration in the wire.

The Generator Principle

Replacing the magnet and plucking the string mechanically, the teacher may apply the discoveries of Joseph Henry and Michael Faraday (1830-31) to predict that an alternating current will be induced in the wire having the same frequency as that of the vibrating wire. If the voltage output so generated in the wire is fed into the low-impedance side of an impedance-matching transformer and the high-impedance side is connected to the vertical plates of an oscilloscope through its built-in, high-gain amplifier, the motions of the vibrating string will be transferred to a spot on the oscilloscope moving in a vertical line at precisely the same frequency.

NOTE: The output transformer from a small discarded table model radio can serve as the matching transformer.

By connecting the internal saw-tooth wave generator to the horizontal plates, the waveform of the vibrating string may be seen on the oscilloscope (Fig. 4). Replacing the internal generator on the horizontal plates with an external variable audio frequency generator will result in various Lissajous patterns from which any frequency of the vibrating string can be determined (Figs. 5-8).

Fig. 4: The wave form produced by a wire vibrating in a magnetic field. The internal saw-tooth wave oscillator is connected to the horizontal plates.

Laws of Vibrating Strings

When the tension is varied (holding length constant), the natural frequency of the vibrating string will also change. Plotted measurements will give a customary parabolic curve (Fig. 9) from which a replot of tension vs. frequency squared—or the square root of tension plotted against frequency—will result in a straight line (Fig. 10). Similarly, measurements of length vs. frequency at constant tension will result in a hyperbolic plot which will rectify itself upon inversion of either variable (Figs. 11 and 12). From these measurements, some of the laws of vibrating strings may be deduced.

Fig. 5: Here a sine wave oscillation from a variable audio-frequency generator is coupled to the horizontal plates. The resulting circular Lissajous pattern represents a 1:1 frequency ratio.

Fig. 6: A 2:1 Lissajous Pattern. The reading of 33 Hz on the audio oscillator (horizontal plates) indicates a vertical (vibrating wire) frequency of 2 x 33 or 66 Hz.

Fig. 7: A 1:3 Lissajous Pattern. The signal generator now reads 198 Hz (horizontal deflection). This indicates a vertical (wire) frequency of 198/3 or 66 Hz, verifying the first measurement.

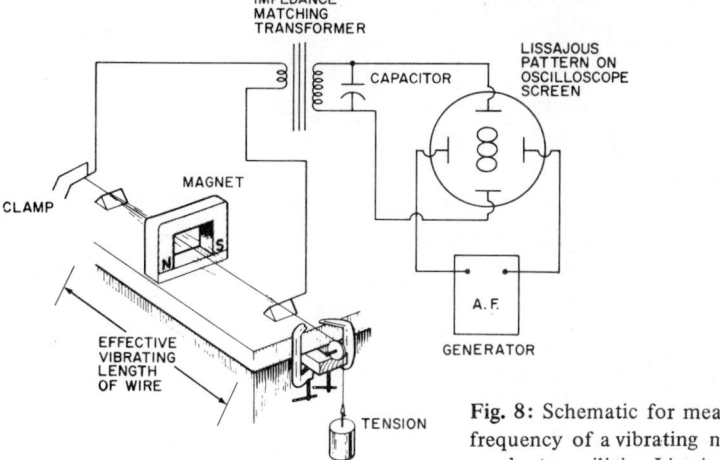

Fig. 8: Schematic for measurement of the frequency of a vibrating nonmagnetic conductor utilizing Lissajous patterns.

Fig. 9: Plot of frequency vs tension for a vibrating wire held at constant length (30 cm).

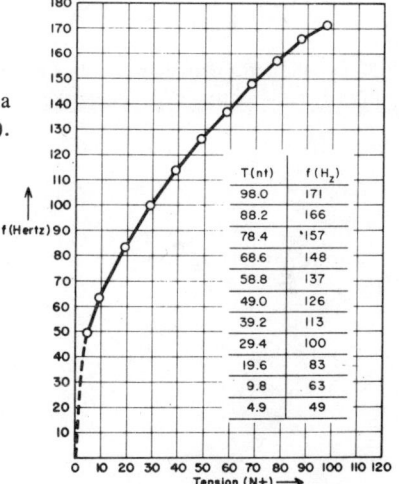

T(nt)	f (H$_z$)
98.0	171
88.2	166
78.4	'157
68.6	148
58.8	137
49.0	126
39.2	113
29.4	100
19.6	83
9.8	63
4.9	49

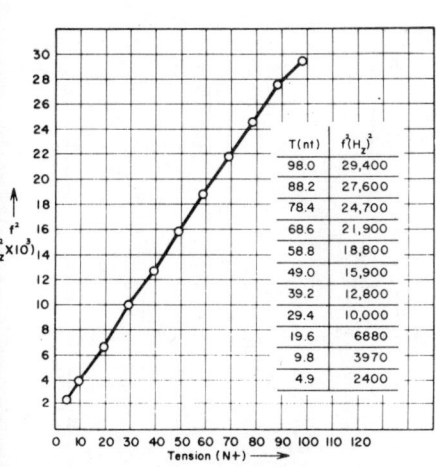

T(nt)	f^2(H$_z$)^2
98.0	29,400
88.2	27,600
78.4	24,700
68.6	21,900
58.8	18,800
49.0	15,900
39.2	12,800
29.4	10,000
19.6	6880
9.8	3970
4.9	2400

Fig. 10: Replot of frequency squared vs tension for the vibrating wire at constant length.

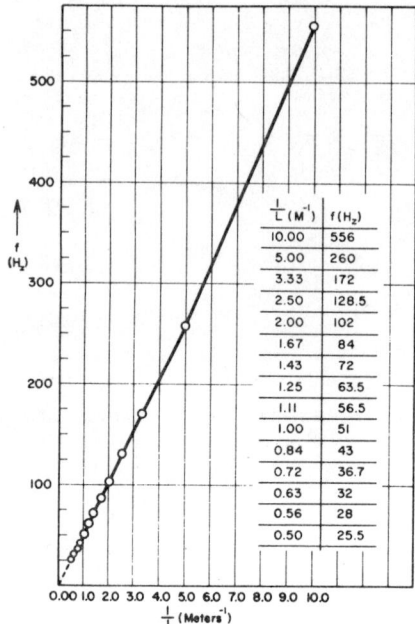

$\frac{1}{L}$ (M^{-1})	f(H$_z$)
10.00	556
5.00	260
3.33	172
2.50	128.5
2.00	102
1.67	84
1.43	72
1.25	63.5
1.11	56.5
1.00	51
0.84	43
0.72	36.7
0.63	32
0.56	28
0.50	25.5

Fig. 11: Plot of frequency vs reciprocal length for a vibrating wire at constant tension.

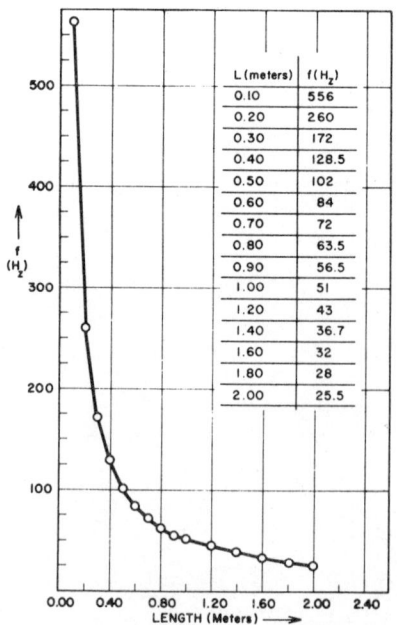

L (meters)	f(H$_z$)
0.10	556
0.20	260
0.30	172
0.40	128.5
0.50	102
0.60	84
0.70	72
0.80	63.5
0.90	56.5
1.00	51
1.20	43
1.40	36.7
1.60	32
1.80	28
2.00	25.5

Fig. 12: Plot of frequency vs vibrating length of wire held at constant tension (98 nt).

5

POLARIZATION
OF LIGHT

A LASER-ASSISTED DEMONSTRATION
OF BREWSTER'S ANGLE

by Thomas G. Cupillari

Keystone Junior College, La Plume, Pennsylvania

In 1812 Sir David Brewster was able to show that light which is reflected from a dielectric surface becomes plane polarized at a particular angle of incidence, the angle being peculiar to the substance investigated. He further proved that the tangent of this angle is equal to the index of refraction of the substance (Fig. 1).

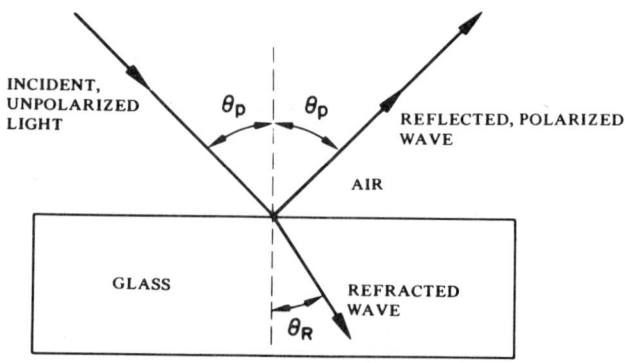

Fig. 1: *For a particular angle of incidence θ_p, the reflected wave is plane polarized. The angle between the reflected and refracted wave is 90°.*

For several reasons this phenomenon is seldom demonstrated in the physics classroom: (1) The reflected, polarized beam from an ordinary incandescent light is too dim to be seen well in a large room. (2) Setting up a high wattage lamp with an adequate condensing lens to concentrate a bright beam of light upon the substance to be investigated takes too much time. (3) This type of light system usually provides scattered and extraneous light which drowns out the reflected beam. To correct this situation requires some system of light baffles.

NOTE: Obviously, few teachers have adequate time to overcome these problems. However, as will be shown here, it is possible to perform the demonstration of Brewster's Angle quickly and easily with any one of the relatively inexpensive gas lasers on the market.[1] The laser eliminates all of the difficulties described above.

(a)   The laser's beam is so intense that the reflected beam is easily seen by everyone, even in a large room with the lights dimmed.

(b)   The laser beam is coherent, so that it needs no lenses and no baffles to limit the scattering of light.

(c)   A self-contained unit, the laser is easily set up.

Materials

To conduct the demonstration, you will need:

Continuous gas laser
Inclined plane similar to Cenco No. 75840
1 Polaroid analyzer
Samples of plate glass, fused quartz, carbon disulfide, water, etc.
Small mirror
Ring stand and several clamps

Procedure (Solids)

With solids, the procedure is as follows:

(1) Refer to Fig. 2 and assemble the apparatus as illustrated. (The mirror allows the reflected beam to be projected upon the wall or screen at a comfortable viewing height, otherwise the beam is reflected high on the wall or on the ceiling.)

CAUTION: NEVER place the laser in such a position that the student can look directly into the optical cavity or into the reflecting beam. The intensity of the beam can damage the eyes.

(2) Arrange the Polaroid analyzer so that it will permit only vertical waves to pass through. (This is done because the polarized, reflected waves are horizontal and cannot pass through the analyzer. When the polarizing angle is reached, all waves are stopped by the analyzer, thus indicating the extinction of the reflected waves.)

(3) Place the sample upon the inclined plane and secure it.

(4) Raise the plane until the reflected light beam becomes weaker and finally extinguishes. The angle of inclination of the plane at which this occurs is the complement of the polarizing angle; i.e., $90° - \phi = \theta$ p.

---

[1]Model 170, $295, Optics Technology, Palo Alto, Calif.; Model 200, $195 and Model 240, $295, University Labs, Berkeley, Calif.; Model LS–32, $285, Electro-Nuclear Laboratories, Inc., Menlo Park, Calif.; Model LAS–2002, $232, Electro Optics Associates, Palo Alto, Calif.

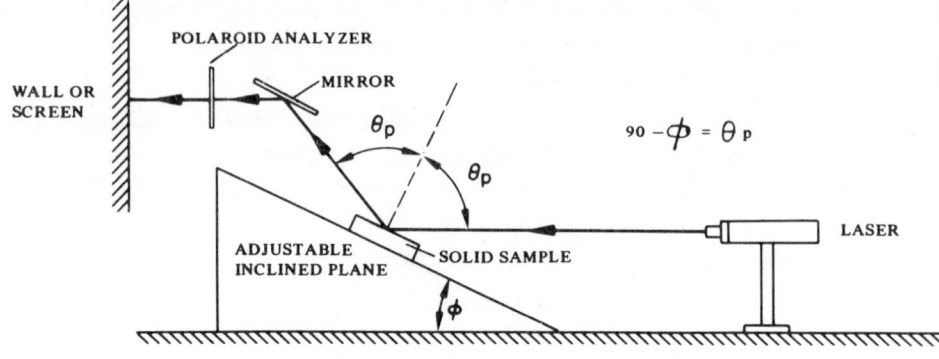

Fig. 2: *Arrangement of the apparatus for reflecting the laser beam off solid samples.*

In the case of common plate glass, the extinction of the reflected beam will occur when the plane is elevated 33.5°, which means that the polarizing angle for this sample is 56.5°. The tangent of θ_p, 56.5°, is 1.52, which is the index of refraction of common plate glass.

Procedure (Liquids)

With liquids, refer to Fig. 3 for setting up the apparatus. Clamp the laser to the inclined plane securely to prevent it from slipping and being damaged. The actual procedure is the same as the procedure followed for solids.

Liquids have an advantage in that with some of them the refracted beam can be seen, and it can be easily shown that the sum of the angles of refraction and the angle of polarization is 90° ($\theta_r + \theta_p = 90°$). This relationship exists because the angle between the reflected beam and the refracted beam is 90° when the beam is polarized.

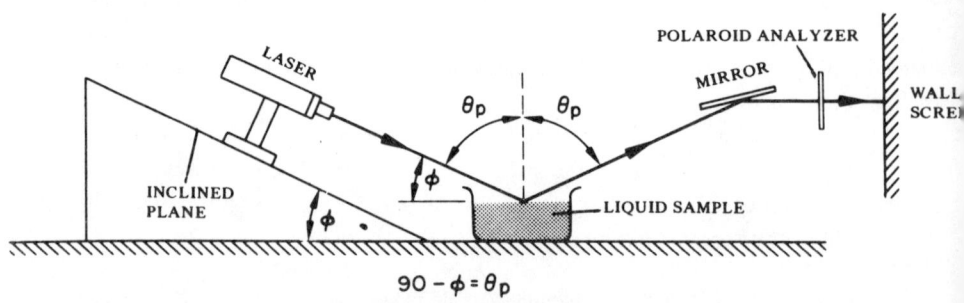

Fig. 3: *Arrangement of apparatus for reflecting the laser beam off liquid samples.*

Typical Values

Some typical values of θ p and n (index of refraction) are given below:

	θ P	n	For λ of
Common plate glass	56.5°	1.52	589 mμ
Fused quartz	55.4°	1.45	
Fluorite	55.0°	1.43	
Carbon disulfide	57.5°	1.63	
Water	53.2°	1.33	
Carbon tetrachloride	55.8°	1.46	

Effective Use

This demonstration can be especially meaningful if it is preceded by (1) an experiment or a demonstration of the determination of the index of refraction of a substance by the ray trace method, and (2) a demonstration of the complete polarization of a light source with the use of two polarizers. Once students have experienced these preparatory exercises, the demonstration of Brewster's Angle of Polarization shows in one experiment that various substances can plane polarize light and that for this substance the angle of polarization is dependent on the index of refraction of the substance.

DEMONSTRATING THE EFFECTS OF POLARIZED LIGHT

by Wendell S. Smith

Kent State University, Stark County Campus,
Canton, Ohio

This demonstration is designed to show the effects produced by polarized light passing through various substances. To be used effectively it should of course be preceded by explanations of the principles and operation of polarized light.

Materials

The following materials are needed to conduct the demonstration:

Slide projector

An extra lens

2 pieces of Polaroid

Suitable materials through which to pass the light

NOTE: One of the materials that can be used to pass light through is easily made by sticking several pieces of cellophane tape to a glass plate in such a way that one, two, three, or more thicknesses are built up at different places on the glass (Fig. 1).

In my own demonstrations, I use a projector with an f3.5 lens and 300-watt bulb. The extra lens has a focal length of 32 cm and is supported by a wooden frame. The Polaroid and the glass plates are supported by ring stand clamps.

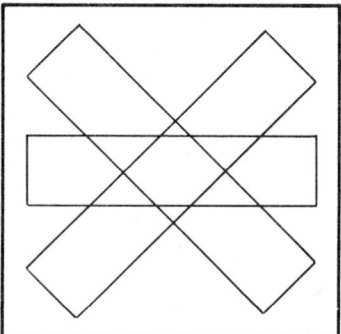

Fig. 1: Cellophane Tape Sample

Procedure

The procedure is as follows:

1. Place the projector in front of a screen at a distance which will produce an image large enough to be viewed comfortably by the class, and focus the projector to produce a clear image on the screen.
2. Place the sample plate into the light from the projector.

NOTE: The distance from the projector is not critical.

3. Position the lens beyond the plate and adjust it so that a clear image of the sample is produced on the screen.
4. Put the Polaroids in the beam of light, one ahead of and the other after the sample. (The one nearest the projector serves as the polarizer; the other serves as the analyzer.) The finished apparatus is shown in Fig. 2.

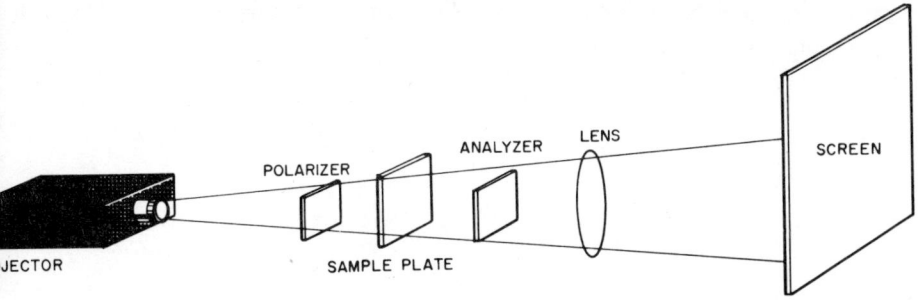

Fig. 2: Demonstration Setup

Polarization of the light will cause areas of different colors to appear in the image on the screen.

Preferential Polarization

To explain the appearance of these different colors, consider a beam of monochromatic light passing through a polarizer and an analyzer in the crossed position. No light will be transmitted. By inserting between them some polarizing material such as a properly cut piece of quartz, some light will be transmitted, and it will be necessary to rotate the analyzer to a new position to again stop the light. If the color of the light is changed, the angle of rotation of the analyzer to stop the passage of the light will also have to be changed.

Thus all wave lengths of light are not polarized to the same degree. For each position of the analyzer, some wave lengths of light will pass and some will not. In the case of white light, a beam polarized under the described conditions will lack some wave lengths, and colors will be seen.

Variations

A variation of this demonstration can be created by replacing the cellophane tape with a blank plate.

IMPORTANT: This blank plate should be *thoroughly* **cleaned. Water should not "bead" on it.**

As the light shines through the blank plate, smear a solution of potassium nitrate over the plate, using a piece of cotton for a swab. When the water evaporates, crystals of potassium nitrate will appear, and the effect produced by polarized light will be seen on the screen.

The demonstration can be enlarged by allowing the students to use various substances in an attempt to determine what substances will polarize light.

Another possibility is to place a sample in the light beam in such a way that stress can be applied. As the stress is applied to the sample, then relieved, a changing pattern of colors should appear on the screen.

6

SPECTROSCOPY

PROJECTS FOR THE
CIGAR BOX SPECTROSCOPE

by Thomas G. Cupillari

Keystone Junior College, La Plume, Pennsylvania

In the April issue of *The Physics Teacher* there appeared an article showing how to construct a simple but functional spectroscope with a cigar box and replica grating.[1] Shortly after the article appeared, I was involved in a program to introduce talented high school seniors to some facets of college physics. The approach was to introduce a concept, develop it, and discuss the theory. Students were then asked to do a project study encompassing the concepts we had discussed.

> **NOTE: When I realized how little the students knew about the spectral properties of light, I decided to assign them a project based upon the cigar box spectroscope. The project met with enthusiasm from the students and proved very effective in developing basic ideas of spectroscopy.**

The purpose of this report is to present the methods used in this project and some of the results. The report first described an improved construction of the instrument which reduces the percentage of error to less than 0.5 per cent yet does not add to its complexity. It then presents several studies for classroom investigation and studies in absorption spectroscopy which the student can perform at home using common household items.

Making a Cigar Box Spectroscope

Materials: To construct the spectroscope, you will need:

Cigar box, the larger the better

Diffraction grating, similar to Edmund Scientific No. 40,272—13,400 lines/inch

3" x 5" index card

---

[1] James Mahoney, "Cigar Box Spectroscope," *The Physics Teacher*, Vol. 5, No. 4, p. 173.

Plastic centimeter ruler or centimeter tape

Razor blade

Construction procedure: Refer to Fig. 1 in carrying out the following directions for making the spectroscope.

1. Cut a ½-inch hole approximately 1 inch in from the front edge of the cigar box.
2. Cut a thin slit from a 3" x 5" index card with a razor blade. (With a little practice, slits in the order of 0.2 to 0.3 mm can be made.)
3. Trim the card to fit and tape it over the hole with the slit centered on the hole.
4. On the opposite end of the box, drill a 1-inch diameter hole on the same center line as the hole cut on the front side of the box.
5. Place the replica grating over this second hole.

NOTE: The 1-inch eyepiece is necessary so that the entire grating surface will be used to diffract light from the source. The resolving power of the grating is in direct proportion to the total number of grating lines that produce the spectrum.

6. Fasten a centimeter tape or a portion of a thin plastic centimeter ruler carefully and securely to the inside front edge of the box.

IMPORTANT: Care should be taken to line up the edge of the ruler as close as possible to the edge of the slit (Fig. 1). This provides a scale for measuring the amount of dispersion of the spectral lines from the entrance slit.

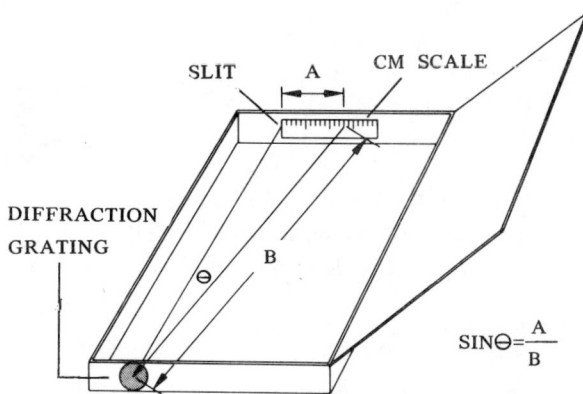

Fig. 1: Construction of the cigar
box spectroscope.

Making measurements: By locating a spectral line on the scale and representing this distance as A and measuring distance B—the distance from the observed line to the center of the grating—we can find $\sin\theta = \frac{A}{B}$. By substituting this value into the equation $\lambda = d \sin\theta$, we can calculate the observed wave length.

In the relation $\lambda = d \sin\theta$, d is the spacing between the lines on the grating. To determine d for the grating, first determine the number of lines per centimeter by dividing the 13,400 lines per inch by 2.54 cm per inch. This gives a result of 5,276 lines per cm. Dividing this value into one $\left[\dfrac{1}{5276 \frac{\text{line}}{\text{cm}}} \right]$ gives a value for d of 1.90×10^{-4} cm. (For the purpose of the example, the grating listed in the Materials is used.)

To make the measurements of the values of A and B easier, a full-sized model of the inside of the spectroscope can be created on a sheet of paper as shown in Fig. 2. This simple device allows measurements to be made with considerable care. The accuracy of wave lengths depends in large part upon the measurement of these two distances, A and B.

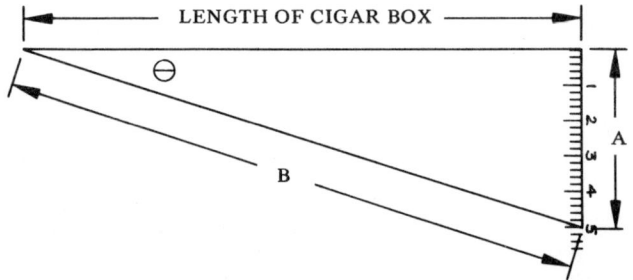

Fig. 2: Scale drawing of spectroscope
and cm scale. (Draw to full
size to determine $\sin\theta$.)

ACCURACY: Table 1 shows the results obtained with this instrument and compares the observed values with the table values for several lines of the Hg spectrum. One can readily see that these values represent errors of less than 0.5 per cent.

Table 1

Line to be measured	Measured value	% error
Hg 4358 Å	4364 Å	0.137
Hg 5461 Å	5444 Å	0.290
Hg 5770-90 Å Doublet	5765 Å	0.396

Studies with the Spectroscope

Bright line spectra: For studies of bright line spectra with the spectroscope, you can use any of the standard high-voltage gas discharge tubes, such as hydrogen, sodium, nitrogen, and cadmium. Fluorescent lighting is another interesting and inexpensive source. This provides a bright line spectrum which is indicative of the Hg vapor in the tube and also shows a continuous spectrum background.

The investigation of fluorescent lighting opens up the discussion as to how fluorescent materials absorb the ultraviolet radiation of the Hg, which is invisible to our eyes, and re-emits this energy in the visible portion of the spectrum. When the Hg spectrum has been previously studied in a discharge tube, the fluorescent tubes also make an interesting "unknown."

Increasing voltage: Use of a 100-watt light bulb in conjunction with a powerstat will allow the student to study the effect of increasing voltage upon the spectrum of the bulb. It is known that as the voltage is increased through the bulb, the temperature of the filament is increased, thus producing a brighter, whiter light. As the voltage is increased from a low level of 35 v up to an operating voltage of 120 v, the spectrum shows red at the low voltage settings and then shifts to yellow, green, blue, and finally violet.

NOTE: The higher voltages are a measure of the greater amount of energy being dissipated through the bulb. This brings up the opportunity to relate increasing energy with decreasing wave length and increasing frequency of light.

Absorption spectra: To this point, the study has centered about the emission type of spectra. Once the class has become acquainted with the theory and explanation of emission spectra, they can be introduced to the study of absorption spectra.

Light from a 100-watt bulb can be passed through a square-base medicine dropper bottle containing substances such as $CuSO_4$ or $NiNO_3$ to produce a continuous spectrum of the bulb with darkened portions. These darkened portions indicate the bands of wave lengths which the substance has absorbed from the continuous spectrum. The pattern and extent of these absorption regions is characteristic of the absorbing medium, just as the bright light spectrum is characteristic of the emitting gas.

Fig. 3 shows a reproduction of the absorption spectra of several common substances. The numbers on this illustration correspond to the scale inside the

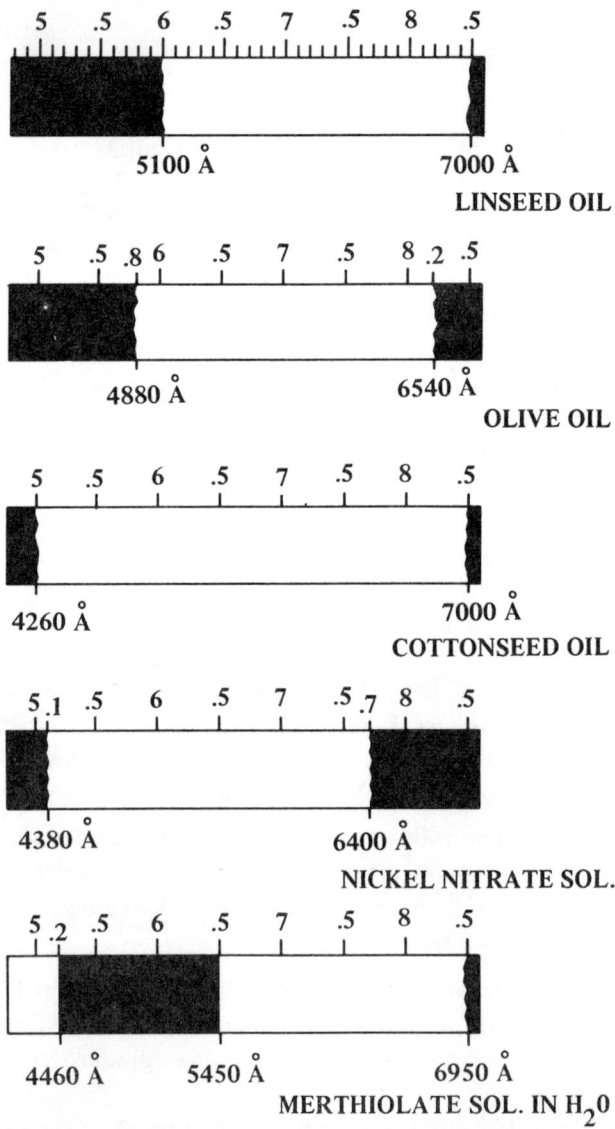

Fig. 3: Reproductions of several absorption spectra.

spectroscope. (The width of the absorption pattern can be determined by calculating the wave length of the edges of the patterns.) The similarity of the three oils is immediately obvious in the similarity of their spectra. This indicates that these oils are similar in their basic molecular structure. It also indicates that they are sufficiently different in their molecular structure that their absorption spectra differ, even if slightly.

HOME STUDY: The student can easily pursue absorption studies at home, using his spectroscope and a small lamp with a 100-watt bulb. Many substances found around the home will serve as interesting mediums, including motor oil, brake fluid, mouth washes, syrups, detergents, and various colored plastics.

Learning value: The preceding suggestions and procedures should provide the student with some rewarding learning experiences. Constructing the spectroscope is a worthwhile experience in itself, giving the student the opportunity to see just how simple this instrument is in its most basic form. And the fact that measurements with such a simple device can be made consistently with errors of less than 0.5 per cent is heartening.

NOTE: With the teacher's leadership, the student can participate in many exciting discoveries: the causes of bright line, continuous, and absorption spectra; the differences between the spectra; and the observation of their many and varied forms.

AN INEXPENSIVE WAY TO DEMONSTRATE PLANCK'S CONSTANT

by Brother James Mahoney

Malden Catholic High School, Malden, Massachusetts

One method of evaluating Planck's constant is from spectroscopic data. But as presented in the PSSC course, the experimental method is involved and expensive. The cost per student group, including a student spectrometer, a power supply, and hydrogen and mercury spectrum tubes is about $65.

This cost can be greatly reduced—with a loss of significant figures in the calculation of Planck's constant, but with a gain in simplification in the experimental technique.

Cigar Box Spectroscope

In place of a spectroscope, which must be calibrated with a mercury discharge tube, students can use a cigar box spectroscope like that illustrated in Fig. 1.[1] The distances x and l can be easily measured and the wave length of the spectral lines of hydrogen can be calculated from the equation: $\lambda = \frac{x}{l} \cdot d$, where d is the distance between lines on the diffraction grating.

Diffraction gratings with 13,400 lines per inch (d equal to 1.9×10^{-6} m) can be bought for a very low price.[2] The hydrogen spectral lines corresponding to 6.6×10^{-7} m (red), 4.9×10^{-7} m (blue-green), and 4.3×10^{-7} m (violet) are easily seen and measured. Students with good eyesight can also see the 4.0×10^{-7} m (deep violet) spectral line.

Two spectrum-tube power supplies and two hydrogen spectrum tubes are adequate for an entire class. One power supply and spectrum tube can be set up at each end of the room, and with their cigar box spectroscopes, students can gather around either of the spectrum tubes.

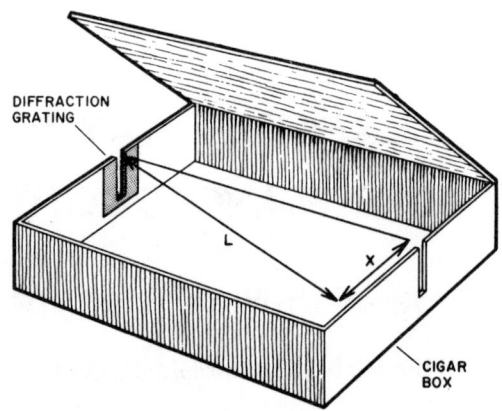

Fig 1. Cigar box with slits cut at opposite ends. Diffraction grating is glued over one slit.

[1]"Cigar Box Spectroscope," *The Physics Teacher,* April 1967, p. 173.

[2]Edmund Scientific Company, Barrington, New Jersey.

Calculations

Since the visible spectral lines of hydrogen are in the Balmer series, students know that the final energy state of the excitation emission is in the $n=2$ energy state. They have learned that:[3]

$$E_n = \frac{-2\pi^2 k^2 m}{h^2 n^2} \qquad k = 2.3 \times 10^{-28} \frac{\text{newton-m}^2}{(\text{elem ch})^2}$$

So,

$$E_{final} = \frac{-2\pi^2 k^2 m}{h^2 n^2_f} \qquad m = 9.1 \times 10^{-31} \text{ kg}$$

and

$$E_{initial} = \frac{-2\pi^2 k^2 m}{h^2 n^2_i} .$$

The energy changes are illustrated in Fig. 2.

Fig 2. Some Balmer series energy transitions.

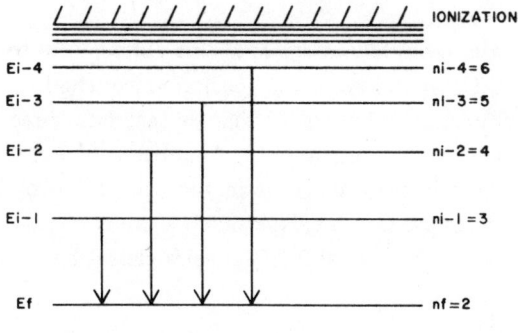

Since a spectral line is emitted with a frequency, γ,

$$h\gamma = E_i - E_f = \frac{2\pi^2 k^2 m}{h^2} \left(\frac{1}{n^2_f} - \frac{1}{n^2_i} \right).$$

For the Balmer series, we get:

$$\gamma = \frac{2\pi^2 k^2 m}{h^3} \left(\frac{1}{2^2} - \frac{1}{n^2_i} \right)$$

---

[3]The factor q^4 is omitted from the equation, but the units of $(1\,\text{e.c.})^4$ are necessary for dimensional analysis to yield units of joule-sec for Planck's constant.

And, since $\gamma = c/\lambda$,

$$1/\lambda = \frac{2\pi^2 k^2 m}{c h^3} \left(\frac{1}{2^2} - \frac{1}{n_i^2} \right).$$

To determine which initial quantum numbers correspond to the measured wave lengths, we observe from the last equation that:

$$\frac{1/\lambda_1}{1/\lambda_2} = \frac{\lambda_2}{\lambda_1} = \frac{\left(\dfrac{1}{2^2} - \dfrac{1}{n_{i-1}^2} \right)}{\left(\dfrac{1}{2^2} - \dfrac{1}{n_{i-2}^2} \right)}.$$

Students can calculate the ratio of any two of the wave lengths measured in the experiment; and, by trial and error, they can calculate which values of n_1 will give them an equal ratio.

For example:

$$1.35 = \frac{\lambda_2}{\lambda_1} = \frac{6.6 \times 10^{-7} \text{ m}}{4.9 \times 10^{-7} \text{ m}} = \frac{\dfrac{1}{2^2} - \dfrac{1}{4^2}}{\dfrac{1}{2^2} - \dfrac{1}{3^2}} = 1.35.$$

Therefore, the 6.6×10^{-7} m spectral line corresponds to a quantum jump from n=3 to n=2. Other ratios for the spectral wave lengths can be evaluated in the same way. Students will then be able to calculate three or four values of Planck's constant with an accuracy to two significant figures. Of course, the fact that one has to extract the cube root in the calculation of Planck's constant reduces the percentage error in the experiment significantly.

Results obtained by several PSSC physics classes last year are listed in the following table:

Value of Planck's Constant ($\times 10^{-34}$ J-sec.)	No. of Student Calculations	Per Cent Error
7.0	2	6.1%
6.9	4	3.0%
6.7	15	1.5%
6.6	62	—
6.5	64	1.5%
6.4	26	3.0%
6.3	1	6.1%

NOTE: Because of its simplicity in measuring the hydrogen spectrum line wave lengths and the low cost of the apparatus used, I believe this method of demonstrating Planck's constant could be advantageous for many situations.

DETERMINING THE WAVE LENGTH
OF SODIUM LIGHT WITH
HOMEMADE EQUIPMENT

by Loren G. Kilmer

Reed City High School and Ferris State College, Reed City, Michigan

Motivated primarily by the eye strain which follows a series of readings on the conventional Michelson apparatus used in measuring the wave length of monochromatic light, I devised a student-built apparatus for accomplishing essentially the same purpose. (See Fig. 1.) It is built and used as follows:

The basic part of the apparatus is a light-tight wooden box, approximately 18" x 10" x 8", constructed with a loose-fitting, recessed cover. A large hole in the removable cover provides an observation post. An old swimming mask, tacked to a wedge-shaped block and nailed over the opening, serves both to exclude light when the device is being used and to steady the head of the observer. A small hole, about ¾" in diameter at one end of the box serves to admit the yellow light of a sodium flame.

Two dowel rods, 5/8" in diameter and inserted into 5/8" holes for a tight friction fit, provide support for a small platform and a hand magnifying glass. The glass is fastened to the top dowel rod, the platform to the lower. (See Fig. 2.)

Two flat glass plates, easily obtainable from any scientific supply house, are taped tightly together at one end. At the other end they are separated by a narrow strip of onion skin paper, whose thickness has been previously determined by a micrometer. This second end is then taped together. With a glass cutter, or piece of quartz, two transverse lines are scribed on one plate, approximately in the center, and 1 cm apart. (See Fig. 3.)

The apparatus works as follows: The prepared plates are fastened with rubber bands to the platform on the lower dowel rod. With the glass in position, the cover is placed on the box, and the light started. The sodium flame is supplied by a Bunsen burner wrapped with an asbestos sleeve soaked in NaCl solution. After the viewer's eyes become adjusted to the dim light, interference bands may be observed on the air wedge. By careful adjustment of the wedge and magnifier (as can be seen in the illustrations, the dowel rods are provided

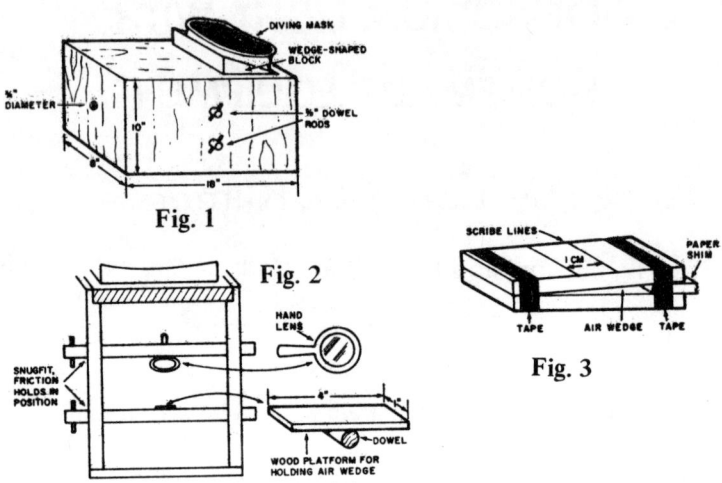

Fig. 1

Fig. 2

Fig. 3

with "handles" for adjustments), a student can easily count the number of dark bands in the 1-cm section of the wedge.

Calculating the Wave Length

As can be seen from Fig. 3, the air wedge has two reflecting surfaces to consider, and these are at a very slight (microscopic) angle to each other. The length of both plates is known, as is the thickness of the paper shim. Reducing this to the rise to the run over a 1-cm range, and dividing by the number of destructive interference bands, gives us, when the wave theory is applied, a means of calculating the wave length of the sodium flame.

Thus, in the air wedge, we have essentially a series of resonance tubes that differ in length from one another by an infinitesimal amount. A wave front entering the top of the air wedge is partially reflected and partially transmitted. This latter portion is similarly reflected and transmitted as it strikes the bottom of the wedge and the more dense upper surface of the bottom glass slide.

Resonance experiments with both sound, water waves, and wave machines show us that a wave is reflected 180° out of phase with the incident wave when striking a barrier more dense than the transmitting medium, while those striking a less dense barrier are partially reflected in an in-phase position. Thus, at the point where the thickness of the air wedge is in multiples of ½, destructive interference of the reflected light will occur.

Example: Assume a shim thickness of 1×10^{-3} cm and a plate length of 10 cm. We have an elongated right triangle whose hypotenuse is very nearly the same length as the base, and whose rise or altitude is 10^{-3} cm (Fig. 4).

Over the 1-cm range in which the bands are counted, we have a similar triangle as follows (Fig. 5):

Suppose 20 bands are counted. For each band another triangle exists, whose dimensions are as shown in Fig. 6:

As already explained, this air-wedge thickness should be equal to ½ to produce destructive interference, so therefore: $2 \times (5 \times 10^{-6})$ cm should equal the wave length of the light being considered. (Figures used are for illustration only.)

Fig. 7 illustrates the wave reflected 108° out of phase.

Using this device, trial runs over a period of years results in $\pm$ 30 Å of accepted values for sodium light.

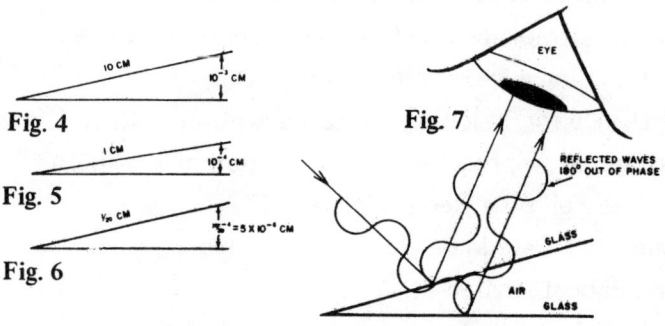

Fig. 4

Fig. 5

Fig. 6

Fig. 7

7

MOTION AT THE EARTH'S SURFACE

HOW FAST CAN YOU THROW A BALL?

by Michael J. Bernstein

Lafayette High School, Brooklyn, New York

The class exercise I use to answer the title question meets three important criteria of a good demonstration: scale (the bigger the better), simplicity, and relevance to the interests and experiences of students. In this excercise, which serves as an interesting two-period mechanics review, students use an apparatus I designed and built to convert the kinetic energy of a thrown ball of clay into the gravitational potential energy of a ball-pendulum system.

Some of the concepts and skills it covers include:

— Conservation of momentum
— Conservation of mechanical energy (gravitational potential and kinetic energy)
— Measurement of mass
— Center of mass
— The plotting and use of graphs
— First law of thermodynamics

NOTE: Deformation of the clay constitutes work on the clay, and the temperature rises.

— Vectors (Consider the effect of not throwing the ball horizontally.)
— Practice in problem-solving, working with units, and using scientific notation

Materials and Apparatus

To conduct the exercise, you will need:

2" x 4" beam 6' to 8' long
Plywood disc, ½" thick, 6" diameter
Nails, 1" finishing
Plastic or cardboard safety cover (anything which will cover the nails will do)
Hammer, file
15' thin string (Nylon fish line is best.)
Loop of cord

2 eye-screws
2 hooks
Nonhardening clay
1 piece 2" x 4" x 24" wood
1 piece 2" x 4" x 18" wood
2 pieces ½" x 18" x 13" plywood

Make a pendulum about 6 to 8 feet in length from a 2 x 4. (The length is chosen so that the bottom of the pendulum is about 5 feet from the floor or shoulder level.) Attach—screw and glue—a plywood disc containing a number of protruding nails with filed off points to the bottom of the pendulum (Fig. 1).

Fig. 1

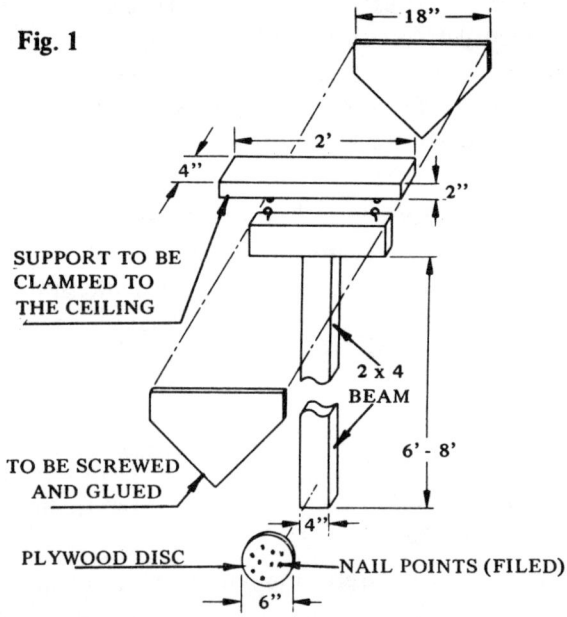

SUPPORT TO BE
CLAMPED TO
THE CEILING

2 x 4
BEAM

6' - 8'

TO BE SCREWED
AND GLUED

PLYWOOD DISC — NAIL POINTS (FILED)

CAUTION: Place a safety cover over the nails when the apparatus is not in use.

Before hanging the pendulum from the 2 x 4 support, which has been previously clamped or bolted to the ceiling, have a student determine the mass of the pendulum on a two-pan balance. Have a second student determine the center of mass by simply balancing the pendulum from a loop of cord.

Next, attach a thin string from the bottom of the pendulum to the string holder, which should be clamped at any convenient location. The string holder (a burette clamp works well) will allow the string to pass through freely as the pendulum swings up. When the pendulum stops and then returns, the string will cease to move through the holder, thus registering the maximum height attained by the pendulum.

Demonstration Procedure

Select a thrower and a measuring team of three students to carry out the following class demonstration:

Instruct the thrower to take a piece of soft, nonhardening clay about the size of a baseball, determine the mass of the clay ball on a two-pan balance, then throw the ball at the disc.

NOTE: I make it a point never to demonstrate this myself because, unlike my students, I usually miss.

The pendulum will swing several feet. The clay must remain in contact with the disc at least to the top of the swing. Then the measuring team should go into action.

Have one student push the pendulum until the string is taut, thus replacing the pendulum to its position of maximum height. Direct another student to hold the cord at the holder, preventing any more string from being pulled through. The third student should measure the gain in height of the center of mass and the gain in height of the clay.

Calculations

The initial velocity of the ball may now be determined by applying two of nature's basic laws: conservation of momentum and conservation of mechanical energy.

(1) The momentum of the ball and the pendulum before contact equals the momentum of the ball and pendulum after contact:

$$m_b v_b + m_p v_{p_i} = m_b v_{b_f} + m_p v_p,$$

where: m_p is the mass of the pendulum.

m_b is the mass of the clay ball.

h is the increase in height of the center of mass of the pendulum.

g is the acceleration due to gravity (9.8 m/sec^2).

v_b is the speed of the thrown ball.

v_p is the speed of the center of mass of the pendulum just after the clay hits it.

v_{b_f} is the speed of the clay moving with the pendulum.

v_{p_i} is the speed of the center of mass of the pendulum before the clay hits it.

(2) Since the pendulum is initially at rest $(v/p_i = 0)$, its momentum before being struck by the clay ball is zero. Thus, the momentum of the clay is converted into the momentum of the clay-pendulum system:

$$m_b v_b = m_b v_{b_f} + m_p v_p.$$

If the mass of the clay is kept small (about 100g), we can neglect the final momentum of the clay ball. For a 5-kg pendulum, the error introduced by this approximation is about 3 per cent. Thus:

$$m_b v_b = m_p v_p, \quad v_p = \frac{m_b v_b}{m_p}.$$

(3) The kinetic energy of the clay plus pendulum after impact is converted into gravitational potential energy:

$$\tfrac{1}{2} \not m_R (v_p)^2 = \not m_R gh$$

$$v_p = \sqrt{2gh}$$

(4) Equating the two expressions for v_p, we obtain:

$$\frac{m_b v_b}{m_p} = \sqrt{2gh}, \qquad \boxed{v_b = \frac{m_p \sqrt{2gh}}{m_b}}$$

REFERENCE: For a complete analysis of the main principles involved in this demonstration, see *University Physics,* **2nd ed., by Sears and Zemansky (Reading, Mass.: Addison-Wesley Pub. Co.), Sections 8-3, pp. 148-152, "Elastic and Inelastic Collisions," and 11-3, pp. 205-207, "Center of Oscillation."**

To increase student participation, have a student plot a graph of speed vs h on the stage of an overhead projector. Once several points have been obtained the curve is drawn. Providing that the mass of the thrown ball is kept constant, i then becomes possible to ascertain very quickly the speed of many throws by measuring h and then simply reading v off the graph.

BULLET BALLISTICS:
A TEACHER LAB DEMONSTRATION

by Lawrence B. Ryan

Beaver River Central School, Beaver Falls, New York

A common problem found in high school physics texts involves firing a bullet into a block of wood and, knowing the masses of block and bullet and the final velocity of the block, employing the Conservation of Momentum to calculate the initial velocity of the bullet.

> NOTE: Since my students often asked whether such a process is "practical," I devised the following lab demonstration. In addition to illustrating a specific problem, it provides a "tour de force" of dynamics.

In this demonstration, the pellet from a .177 or .22 caliber CO_2 gas pistol is fired by the teacher into a lump of clay attached to a "Hall" cart. The resulting speed of the cart is measured by a tape timer. Using the tape measurements, students then develop a series of calculations which nicely illustrate:

(a) The Conservation of Momentum: The momentum lost by the pellet is gained by the cart.

(b) Newton's Second Law: The force acting on each object can be calculated by knowing the mass and acceleration in each case.

1:

aratus Setup

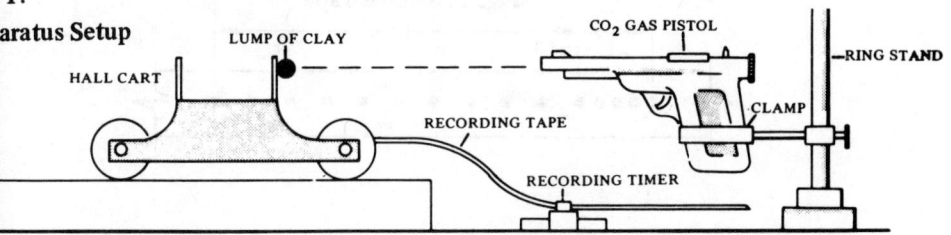

LUMP OF CLAY   CO₂ GAS PISTOL   —RING STAND

HALL CART   RECORDING TAPE   CLAMP

RECORDING TIMER

(c)   Newton's Third Law: For every action (pellet exerting a force on the cart), there is an equal but opposite reaction (force exerted by the cart on the pellet).

Apparatus

To conduct the demonstration, the teacher will need:

.177 or .22 cal. CO_2 gas pistol (single-shot type preferred)
Support rods and clamps
Hall-type lab cart (mass of 400 to 500 grams)
Modeling clay
Recording timer and timer tape

WARNING: THIS APPARATUS IS DANGEROUS IN INEXPERI-ENCED AND OVERENTHUSIASTIC HANDS. IT SHOULD BE SET UP AND USED ONLY BY THE TEACHER.

Demonstration Procedure

Assemble the apparatus as shown in Fig. 1, being careful to securely clamp the pistol with the barrel horizontal. Shape about 100 grams of clay into a sphere and press it to the rear of the cart to absorb the impact of the .177 cal. pellet. A larger mass of clay will be necessary with the .22 cal. pellet.

Carefully align the apparatus, placing the clay directly in the line of fire.

Load and cock the pistol, start the timer, and fire.

NOTE: Enough "firings" may be made to provide each lab team with a tape to analyze.

Analysis and Interpretation

Examination of the tape (Fig. 2) will show two distinct sections: A period of acceleration (a) during which the pellet and the cart were interacting with each other, and a period of constant velocity (b) after the interaction has occurred and when the cart has acquired momentum.

Fig. 2: Sample Tape

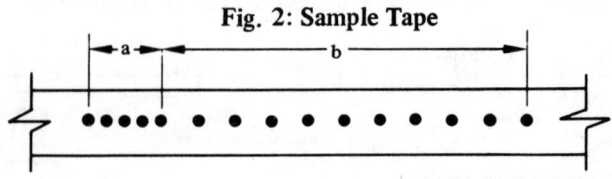

To illustrate what can be done with this recording, following are data and parameters from an actual "run" together with the appropriate calculations:

Mass of cart + clay = M_c = 450 g
Mass of pellet = m_p = 0.5 g
Timer rate = 40 beats/second
Length of interaction = $\triangle t$ = 4 beats = 0.1 sec
After interaction, cart moves 3.5 cm in 10 beats

Calculations

(1) The most obvious first point is to calculate the velocity of the cart, V_c.

$$V_c = \frac{3.5 \text{ cm}}{10/40 \text{ sec}} = 14 \text{ cm/sec} =$$

$$1.4 \times 10^{-1} \text{ m/sec} = V_c .$$

(2) Since the cart was originally at rest, it has now acquired momentum. This momentum, P_c, equals $M_c \times V_c$:

$$P_c = 4.5 \times 10^{-1} \text{ kg} \times 1.4 \times 10^{-1} \text{ m/sec} =$$

$$6.3 \times 10^{-2} \text{ kg - m/sec.}$$

(3) Since the cart picked up speed, it underwent an acceleration, A_c:

$$A_c = V_c / \triangle t = \frac{1.4 \times 10^{-1} \text{ m/sec}}{10^{-1} \text{ sec}} =$$

$$1.4 \text{ m/sec.}^2$$

(4) And, since the cart underwent an acceleration, Newton's Second Law tells us that there must have been a *force* acting *on the cart*. This force, F_c, is produced *by the pellet* and has a value:

$$F_c = M_c \times A_c = 4.5 \times 10^{-1} \text{ kg} \times 1.4 \text{m/sec}^2 =$$

$$6.3 \times 10^{-1} \text{ Nt.}$$

(5) Now let's turn our attention to the pellet. Applying the Conservation of Momentum, we can calculate the velocity of the pellet before impact, if we assume that the momentum of the pellet beforehand, P_p, is equal to the momentum of the cart afterwards; that is, $P_p = P_c$.

PELLET MASS: Of course calculations of momentum after the interaction should include the pellet's mass. But, since this mass is so small in comparison to the mass of the cart, it is well within the limits of error of this exercise to ignore the mass of the pellet.

Therefore, the velocity of the pellet, v_p, is given by:

$$v_p = \frac{M_c \times V_c}{m_p} =$$

$$\frac{4.5 \times 10^{-1} \text{ kg} \times 1.4 \times 10^{-1} \text{ m/sec}}{5 \times 10^{-4} \text{ kg}}$$

$$v_p = 1.26 \times 10^2 \text{ m/sec.}$$

This value is consistent with the manufacturer's published data for the muzzle velocity of the pellet.

(6) The pellet slows down, undergoing a change in speed, $= -\triangle v_p$, where:

$$-\triangle v_p = -(1.26 \times 10^2 \text{ m/sec} - 0.014 \times 10^2 \text{ m/sec}) \cong$$

$$-1.25 \times 10^2 \text{ m/sec.}$$

There is, therefore, a change in the momentum of the pellet, $\triangle P_p$, given by:

$$\triangle P_p = m_p \times \triangle v_p = 5 \times 10^{-4} \text{ kg} \times (-)$$

$$1.25 \times 10^2 \text{ m/sec} =$$

$$\triangle P_p = -6.25 \times 10^{-2} \text{ kg - m/sec.}$$

This is almost exactly the momentum acquired by the cart.

(7) We can now calculate the pellet's acceleration, a_c, where:

$$a_c = \frac{\triangle v_p}{\triangle t} = \frac{-1.25 \times 10^2 \text{ m/sec}}{10^{-1} \text{ sec}} =$$

$$-1.25 \times 10^3 \text{ m/sec.}^2$$

(8) To produce this acceleration, of course, a force is required. Since this force must act *on the pellet,* it must be produced *by the cart.* This force, F_p, has a magnitude given by:

$$F_p = m_p \times a_p$$

$$F_p = 5 \times 10^{-4} \text{ kg} \times -1.25 \times 10^3 \text{ m/sec}^2 =$$

$$-6.25 \times 10^{-1} \text{ Nt.}$$

NOTE: The force is equal to, but in the opposite direction from, the force of the pellet on the cart, as it must be according to Newton's Third Law ($F_p = -F_c$).

APPARATUS FOR DEMONSTRATING NEWTON'S SECOND LAW

by Herbert H. Gottlieb

Martin Van Buren High School, Queens Village, New York

According to Newton's Second Law ($F = ma$), the acceleration of a body is directly proportional to the unbalanced force acting upon it and inversely proportional to the mass of the body. Usually it is impossible to demonstrate this relationship without a considerable amount of manipulative skill on the part of the teacher, as well as an expensive and sophisticated timing apparatus.

NOTE: The apparatus described here does not require any timing devices and is thus easily operated by both teacher and students. Furthermore, it can be constructed in a few minutes from pieces of scrap wood.

Apparatus Construction

Directions for making the apparatus are as follows:

(a) Obtain a wooden board, approximately 8" x 10". Cut out an area in one end of the board large enough on which to stand a paper milk container, as shown in Fig. 1.

(b) Place wooden dowel sticks or large nails at the sides of the cut-out area of the board to act as launching posts.

(c) For the propelling force, stretch one or more rubber bands over the launching posts and press the projectile into the cut-out.

(d) Place known masses inside the projectile or remove them as desired.

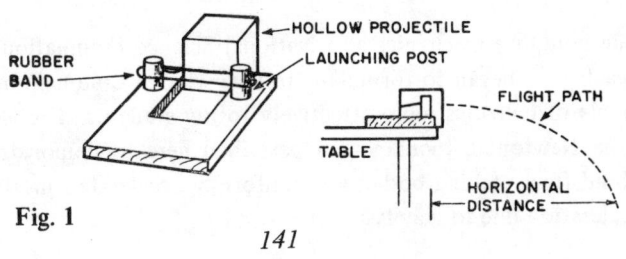

Fig. 1

Operation

Following are procedures for using the apparatus:

(1) Place the projectile in the launcher and measure the horizontal distance it travels when propelled with one rubber band. Repeat this for several trials to determine the average distance.

(2) Launch the same projectile with two rubber bands on the launching posts and record the average distance after several trials. When the force is doubled, students will observe that the horizontal acceleration is greater and the distance of travel will also be greater. Repeat this with three and possibly four rubber bands.

(3) Using a constant number of rubber bands for the propelling force, add known masses to the projectile and record the distances that are associated with each mass. An inverse relationship will be observed; that is, the distance traveled will decrease as the mass of the projectile is increased.

Follow-Up

For enrichment work, have the students graph the launcher force (measured in rubber band units) vs the horizontal distance traveled. Good students might be able to explain why the graph is *not* a straight line. The reason is that the projectile is in the launcher for comparatively shorter times as more rubber bands are used. As the mass is increased, the inertia of the projectile effectively lengthens the acceleration interval. Thus, the horizontal distance will be directly proportional to the *square root* of the force and inversely proportional to the *square root* of the mass.

BUILDING AND USING A
NEWTONIAN MONKEY GUN

by Julius Joseph Breit
Toll Gate High School, Warwick, Rhode Island

While studying mechanics at a National Science Foundation Institute for physics teachers, I began to formulate an idea which would help me present the fundamentals of ballistics more effectively to my students. The idea led to my building the Newtonian monkey gun described here to demonstrate the principles behind freely falling bodies and uniformly accelerated motion as applied to the acceleration due to gravity.

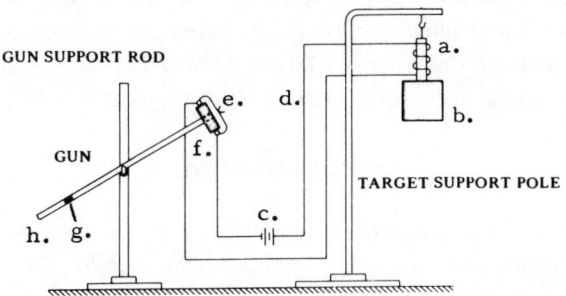

Fig. 1: Newtonian Monkey Gun

a. SIMPLE
ELECTROMAGNET

PLASTIC INSULATOR f.

b. TARGET (LARGE CAN)

c. D.C. POWER SOURCE
(LOW VOLTAGE)

PROJECTILE   g.
d. LEAD WIRE   (BRASS SLUG MACHINED
TO FIT TUBE)

e. FINE COPPER WIRE
STRIPPED OF INSULATION

TUBING ABOUT 2½' LONG h.

COST: Although a similar gun may be purchased from major scientific equipment houses, the gun I built—shown in Fig. 1—can be made for practically no cost. The necessary parts can be found in any high school physics laboratory or metals shop.

This laboratory built model has proven effective in giving life to the topic of projectile motion. It achieves the objective with a very significant percentage of success, and always rouses student enthusiasm. The collision of projectile and target provides an excellent stimulus for discussing Newton's motion formulas as applied to the acceleration of gravity. Also, the projectile travels at a slow enough rate to enable students to actually see its parabolic trajectory.

Construction Notes

The gun is placed approximately 10 to 15 feet from the target, which is hung on a support pole and can be as high as the ceiling in the room permits. Essentially, the tubing serves as a "blow gun." When the circuit is energized, the target is held in place by the electromagnet on a crossbar at the top of the target pole. The gun can thus be aimed directly at the target by sighting the target through the tube.

It is quite important to use fine copper wire which is slightly twisted to make a good electrical contact.

CAUTION: Be sure that the copper wire is attached to the tube by means of a plastic nonconducting medium.

Once the wire has been connected, the projectile can be inserted into the end of the tube. Then, upon blowing through the tube, the force of air will cause the projectile to part the fine copper wires, thus breaking the simple direct current circuit and simultaneously causing the target to begin its free fall.

Student Investigation

Various hypothetical ballistic problems may be discussed preceding an actual experiment with the gun. Students can employ their knowledge of trigonometry and measure the angle of elevation quite easily. Also, they can see how the vertical velocity vector changes from a maximum value when the projectile leaves the end of the tube to a minimum value at the maximum height of trajectory, and then back to a near maximum value upon striking the target.

The experiment can be made more quantitative if the time the projectile is in flight is measured. A rather rough measurement can be made by employing a hand-operated stopwatch. This time measurement, together with a measurement of the range and angle of elevation, will supply a good exercise for the student in the calculation of the projectile's initial velocity. Once this value has been determined, a number of worthwhile computations can be carried out.

EXAMPLE: For instance, these computations could involve the prediction of the height at which the collision between target and projectile will take place.

Equations such as the following may be used in the investigation:

Initial vertical velocity of the projectile: $Viy = Vi \sin\theta$

Horizontal velocity of the projectile: $Vix = Vi \cos\theta$

Range of the projectile: $R = (Vi \cos\theta)t$

Maximum height of the projectile: $h = (Vi \sin\theta) + 1/2\ gt^2$

Where

Vi equals the muzzle velocity
Viy equals the initial vertical component of velocity
Vix equals the initial horizontal component of velocity
θ equals the angle of elevation
t equals time
g equals the accepted value for the acceleration due to gravity

A SIMPLE DEVICE FOR
TEACHING PROJECTILE MOTION

by Edmond Lonsky

Plainfield High School, Plainfield, New Jersey

Here is a simple device I use to help in teaching projectile motion and other areas. You can use either version of a simple "spring gun" that we have made.

Apparatus

In the early design of the MacAlester-Bicknell collision carts, used in the PSSC laboratory program, dowel thrust rods were used in conduit tubes to provide spring bumpers for the carts. The conduit tube on this model of the collision cart can be removed easily by unscrewing a wing nut. Two of these conduit tubes, with one dowel thrust rod, provided our first model of a "spring gun" (called a gun from here on in). Fig. 1 shows how it was accomplished.

Some students, however, encountered some difficulty in handling the gun since (as seen in Fig. 1) the two conduit tubes had to be held together during the exercise. We thus devised a more versatile gun for the exercise. It is made from an 18" length of conduit tube (cut from a 10-foot length of tube purchased in an electrical supply house for $1.50). The tube was drilled with a 3/16" hole about 1/4" from the bottom of the pipe, to accommodate the 1/8" machine screw from the collision cart.

Then, because of the length of the dowel thrust rod, 1/8" holes were drilled

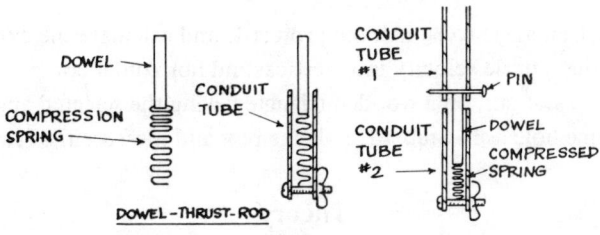

Fig. 1 : Dowel Thrust Rod

at distances of 9", 9 3/4", 10 1/2", 11 1/4", and 12" from the bottom of the tube. (See Fig. 2.)

In addition, wooden triangles were used to rest the gun on in order to get proper angles of projection. A number of them can be cut from pieces of wood ½" to ¾" thick. Useful shapes are 30°, 60°, 90° triangles, and 45°, 45°, 90° triangles (Fig. 3).

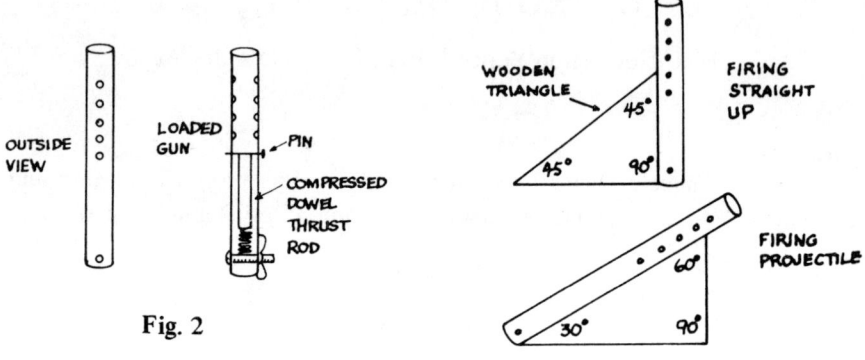

Fig. 2

Fig. 3 : Firing Gun

Experiment

The objective of the experiment is to calculate the range of a projectile and to compare the calculated range with the experimental value. Here is the procedure:

1. Load the gun by using a dowel ramrod to push the dowel thrust rod past a selected hole. Push a nail (pin) through the hole to keep the spring depressed. Drop a projectile (marble, ball-bearing, or piece of chalk) into the barrel.

2. Point the gun straight up (steadied by a wooden triangular stand) and pull out the pin. Determine the maximum height reached by the projectile by means of a marked scale (on wall or on a special piece of board). From this measurement, calculate the muzzle velocity of the gun. [See "Theory" (1.) following.]

3. Select an angle to launch the projectile and calculate the expected range by resolving the muzzle velocity into vertical and horizontal components.

4. Set up the gun on a wooden triangle having the selected angle, cock the gun to the same hole as in Step #1, pull the pin, and measure the range.

Theory

1. Calculate the muzzle velocity of the gun as follows:

d = distance straight up

g = acceleration due to gravity

t = time to reach top of flight

v_m = muzzle velocity

$d = \frac{1}{2} gt^2$ (solve for t)

$v_m = gt$

2. From the muzzle velocity and angle of projection, calculate the range. (See Fig. 4.)

v_v = vertical velocity

v_h = horizontal velocity

t_f = time of flight

$v_v = v_m \sin 30°$

$v_h = v_m \cos 30°$

$t_f = \dfrac{v_h}{g} .2$

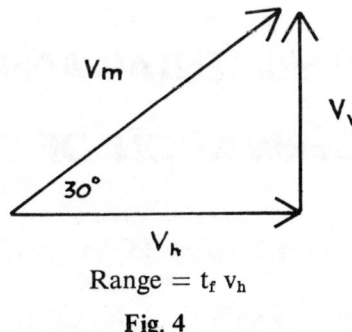

Range $= t_f \, v_h$

Fig. 4

Notes

(a) Measurements here are given in inches because you will probably have your machine shop make the guns.

(b) Later models of the collision carts can be adapted to provide a similar type of apparatus.

(c) The muzzle velocity is calculated from the vertical distance, rather than by time measured directly, because the short time interval introduces error due to reaction time.

(d) The distance calculated in the experiment is from the muzzle of the gun to the same height above the floor on the other end of the trajectory. To use the floor distance introduces an error of about 1 foot in calculated ranges.

Other Uses for the Gun

Once having made your gun, it can be used for a variety of other experiments or demonstrations:

1. A force-compression curve can be plotted for a spring, and the height a given projectile can be shot into the air can be calculated from the spring potential energy stored in the compressed spring.

2. The monkey and hunter demonstration often discussed in projectile motion (shown in the PSSC film) can be shown using this gun.

3. Action-reaction can be shown by suspending the gun from strings and firing it while suspended.

4. A ballistic pendulum experiment can be demonstrated.

5. By using all five holes in the barrel, a plot can be made of height vs V_v , compression of spring.

AN EXPERIMENTAL APPROACH
TO THE DERIVATION OF D = $\frac{1}{2}$AT^2

by David Kutliroff

New Brunswick Senior High School, New Brunswick, New Jersey

Early in the year, we attempt to teach kinematics to our physics classes and arrive at the relationship $D = \frac{1}{2} AT^2$. While this can be done both algebraically and graphically, I believe that kinematics can become more meaningful if this formula were to be experimentally derived and then supplemented by the other treatments. It also provides a means of introducing, early in the course, the experimental method and the analysis of experimental results.

NOTE: The laboratory exercise described here is for two periods, and I have found that it may require another class period for review. The time spent is, I believe, well worth it in terms of a better understanding of the kinematic relationship.

For my experimental approach, I use the bell-clapper timers and tapes designed for the PSSC laboratories on kinematics and dynamics (these may be purchased from many sources; mine come from Macalester Scientific Corp., Cambridge, Mass. 10039). As a matter of fact, this laboratory exercise grew out of an extension of PSSC lab 1-5.

Here is how the equipment works: A weight is attached to the paper tape and is allowed to fall to the floor while the timer (Fig. 1) makes its carbon marks on the tape. (A small C-clamp is a convenient weight to attach to the paper tape.) The tape will then show the characteristic acceleration pattern of dots (Fig. 2). The student is then asked to plot Displacement vs Time. The steps are as follows:

1. For his *time* interval, he can assume that the period of the bell clapper was constant and that one, two, or three claps might constitute one interval of time. He can then lay his tape down along the Y axis of the graph (Fig. 3), rule off the T axis in the regular time intervals, and mark off the points in his graph as shown in line A of Fig. 3.

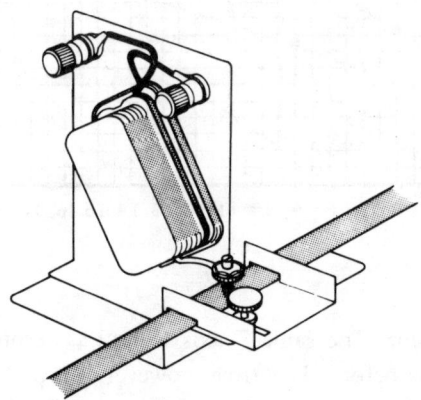

Fig. 1

ALTERNATIVE TIMER: If you don't have a bell-clapper timer, you can make one. See instructions at the end of this article.

2. It is evident from the curve that D is not linearly proportional to T. The curve, however, looks as if it might be parabolic, and this "hunch" can be explored by plotting D vs T^2. This is done on the same graph (or separate identical graph) by leaving the first interval where it is. The dot for the second interval is moved to 4 (2 squared), the dot for the third interval is moved to 9 (3 squared), the dot for the fourth interval is moved to 16, and so on. A line (line B on Fig. 3) connecting these dots appears to be a straight line and, therefore, D is linearly proportional to T^2. The slope of the line (K) is $\dfrac{D}{T^2}$ and thus $D = KT^2$.

3. The experiment can be further extended by suggesting that K may bear some relationship to acceleration. The acceleration is discovered by plotting the

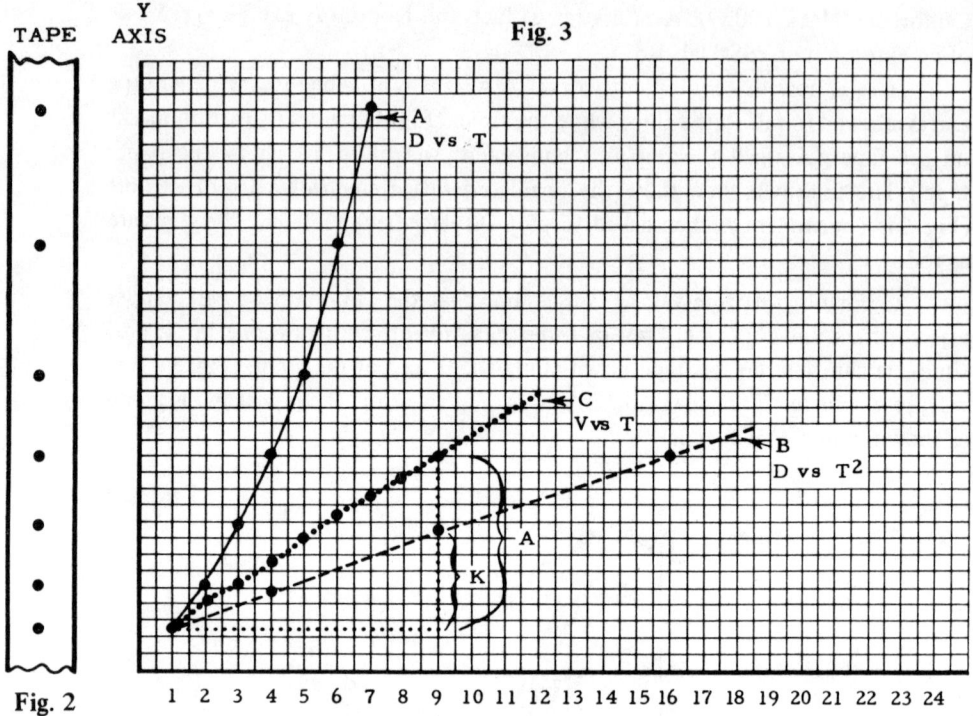

Fig. 2

Fig. 3

Velocity against Time. The same T axis is used as before, and the tape is laid against the Y axis as before. This time, however, it is pulled down each time we plot succeeding displacements to place each succeeding dot on $Y = 0$. By doing this, we plot changes in Distance per unit of Time, rather than Distance in total Time. The resulting straight line (line C on Fig. 3) indicates that V is linearly proportional to T. The slope V/T is the acceleration, and thus $V = AT$. The numerical value of this slope (A) which is in units of V/T or $D/T/T$, can be compared with the numerical value of the slope (K) in the previous line (K is in

dimension of $\frac{D}{T^2}$). It will be then discovered that within experimental error, $K =$

½ A and thus ½ A can be substituted for K in the formula $D = KT^2$, to give $D = \frac{1}{2} AT^2$.

Having thus first derived the formula by experiment, we can then bring in the following algebraic and graphic derivations to supplement the experiment. This gives confidence in the use of mathematics as a technique for predicting physical phenomena.

Algebraic

The derivation can be arrived at by algebraic manipulation of the definitions as follows:

$$\overline{V} = \frac{D}{T}$$ (Average velocity is total change in displacement per interval of time.)

$$A = \frac{Vf}{T}$$ (Acceleration is the total change in velocity per unit of time, and the final velocity is the total change in velocity if the object starts from rest.)

but

$$\overline{V} = \frac{1}{2} Vf$$ (The average velocity is one-half the final velocity if the object starts from rest and accelerates at a constant rate to the final velocity.)

Therefore:

$$D = \overline{V}T$$
$$D = \frac{1}{2} VfT \text{ and } Vf = AT$$
$$D = \frac{1}{2} AT \cdot T = \frac{1}{2} AT^2.$$

Graphic

Another useful presentation is to derive the relationship graphically by graphing V vs T (selecting figures) as V increases regularly with T, as is shown in Fig. 4. (Note comparison with line C, Fig. 3.)

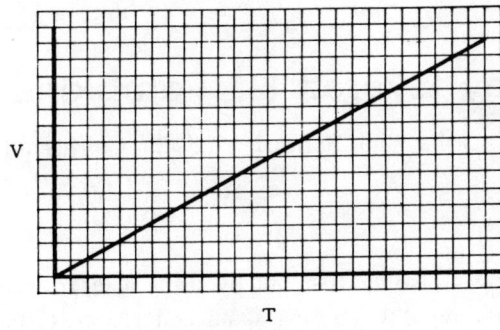

T

Fig. 4

Alternative Timer

Here's how you can make a timer from a doorbell from which the gong has been removed, a microscope slide, some thumb tacks, and wood.

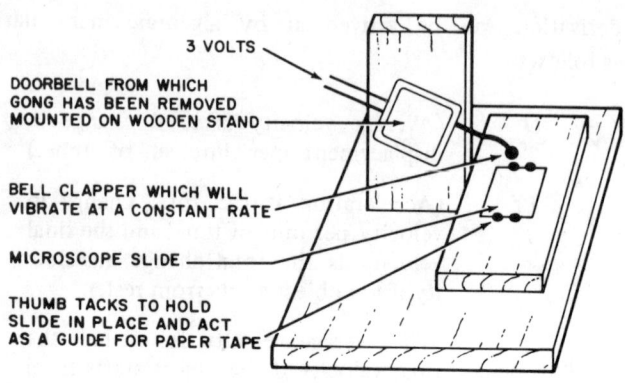

3 VOLTS

DOORBELL FROM WHICH
GONG HAS BEEN REMOVED
MOUNTED ON WOODEN STAND

BELL CLAPPER WHICH WILL
VIBRATE AT A CONSTANT RATE

MICROSCOPE SLIDE

THUMB TACKS TO HOLD
SLIDE IN PLACE AND ACT
AS A GUIDE FOR PAPER TAPE

Fig. 5

Construct the basic timer as shown in Fig. 5 (it can be operated by a pair of flashlight batteries). Then operate as follows: The recording tape (strip of paper about ¼ inch wide) is drawn under the vibrating bell clapper over a hard surface (the glass microscope slide) while a piece of carbon paper is moved around between the bell clapper and the moving tape, in order to present a fresh surface for marking on the paper.

STUDYING FREE FALL ON A RECORD PLAYER

by Everett B. Tompkins

La Jolla High School, La Jolla, California

Here's a way to study free fall by using a record-player turntable (a constant-speed rotating disk), a ring stand and clamp, carbon paper, polar-coordinate graph paper, two balls, and thin thread. Set up the device as shown in Fig. 1. Ball A and ball B are tied together with a length of thread and draped across the fingers of the clamp. They are lined up along a radius of the turntable. With the table turning, the thread is burned and each ball, as it hits, will leave a mark on the graph paper. The angular distance between the marks and the speed of the turntable are used to determine free-fall time.

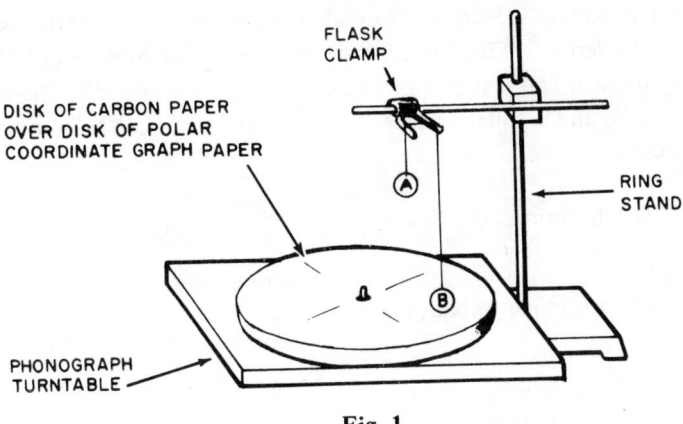

FLASK
CLAMP

DISK OF CARBON PAPER
OVER DISK OF POLAR
COORDINATE GRAPH PAPER

(A)

RING
STAND

PHONOGRAPH
TURNTABLE

Fig. 1

Example One

Ball A is placed 1 foot or more above the turntable, and ball B is barely above the turntable (the drop time of ball B, therefore, is approximately zero). After the drop, the marks on the graph paper are 117° apart. The turntable is rotating at 78 rev/min (or 1.3 rev/sec). The drop time is computed for ball A as follows:

$$\frac{117}{360} \text{ rev} \times \frac{1}{1.3} \text{ sec/rev} = .25 \text{ seconds}.$$

NOTE: If the carbon paper is hard, and it is difficult to see the mark because of the short drop of the ball, you might try gluing a small ball or pointed object to the bottom of ball B to insure seeing the starting point.

Example Two

The height of ball A may be varied to study the time vs. drop distance relationship of two falling objects separated in space. Therefore, raise the clamp so that ball A is still 1 foot above ball B, but ball B is 6 inches from the turntable. When dropped, the impact-mark measurements made by the balls will indicate the time interval between the impact times of the two balls.

CLASS QUESTION: Ask the class: Will the calculated time interval be .25 seconds as before? Let the students argue the point before doing the demonstration.

The time, of course, is less than .25 seconds. Then discover what relationship exists between the separation distance of the two balls (d_s), the height of ball B above the table (d_b), and the time interval between the impact times of the two balls (t_s). Since the formula will not be in most textbooks, it's a good exercise to derive it. First, plot the situation as shown in the graph (Fig. 2). The velocity of each ball is plotted against time on the one graph. The distance ball A drops is d_a and is equal to distance d_b plus distance d_s. The computations follow the graph:

The drop time of ball B is:
$$t_b = \sqrt{\frac{2\, d_b}{g}}$$

The drop time of ball A is:
$$t_a = \sqrt{\frac{2\, d_a}{g}}$$

or, since $d_a = d_b + d_s$:
$$t_a = \sqrt{\frac{2\,(d_b + d_s)}{g}}$$

Now, $t_s = t_a - t_b$, and therefore $t_s = \sqrt{\dfrac{2\,(d_b + d_s)}{g}} - \sqrt{\dfrac{2\, d_b}{g}}$.

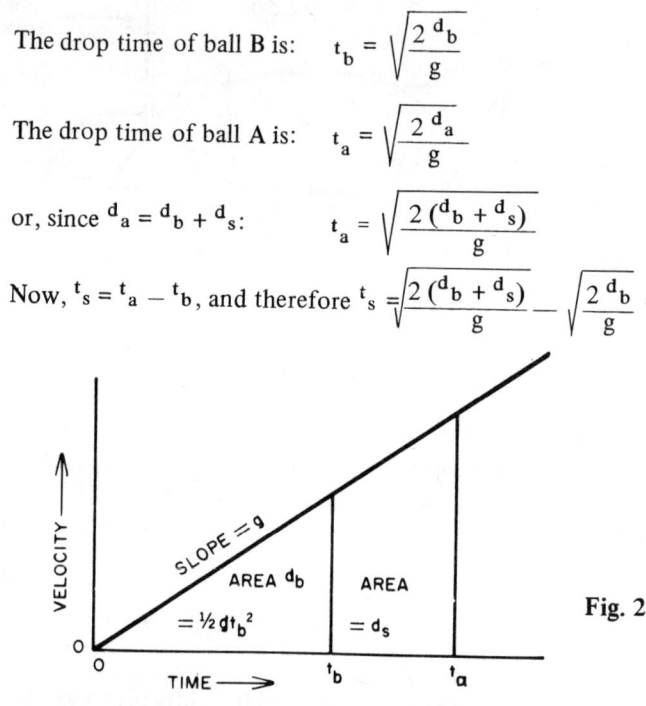

Fig. 2

THE HINGED STICK PARADOX

by Gerrit Zwart

Suffern High School, Suffern, New York

You won't find this one in many texts, and it is a wonderful attention-getter and gets lively discussion under way on the behavior of free and not-so-free falling bodies.

Materials

Obtain a *flat* meter stick, a half-pint milk container, thumb tack, piece of Kleenex, small elastic band, brick (or any other convenient "stop"), and a 5-cent piece.

Fasten the milk container to the stick with the thumb tack (first cutting the container as shown in Fig. 1). Place the elastic band around the meter stick at about the 98-cm mark (its purpose is just to keep the coin from sliding). Place the assembly flat on the table and near the edge, as shown in Fig. 2, with the zero end of the stick against the brick, or stop. The table should be flat and smooth.

Demonstration

Place your right index finger at the edge of the table at a point marking the line of the center of the milk container. Now raise the stick with the left hand as shown in Fig. 3 so that the loosely placed nickel is exactly above the spot where the container was on the table (keep your right index finger on the table to help line up the coin). Let go of the stick evenly so that it strikes the table squarely. After a little practice, you will find that the nickel lands in the *paper container every time!* If it pops out, stuff a piece of fluffed-up Kleenex in the bottom of the container.

> **SOLUTION: The phenomenon is that the edge of a hinged stick, falling, accelerates at a greater rate than that of normal gravitational acceleration. (Obviously, the end of the stick next to the stop accelerates at a rate less than "g".) Offer a prize, or a compliment, to the student giving the best explanation.**

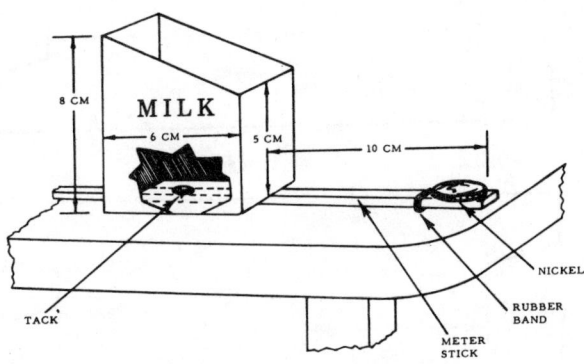

Fig. 1

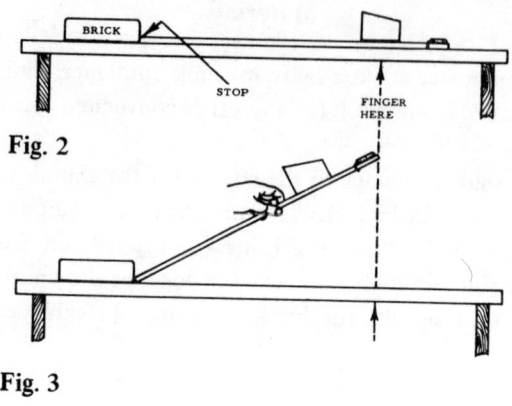

BRICK

STOP

FINGER HERE

Fig. 2

Fig. 3

A DRIPPING FAUCET ANALOG

by Terrence P. Toepker

Xavier University, Cincinnati, Ohio

    A simple yet very useful device for demonstrating the kinematics of freely falling bodies can be made from a piece of string, about eight pieces of split shot, and an appropriately cut piece of cardboard (Fig. 1a). The device is easily tested

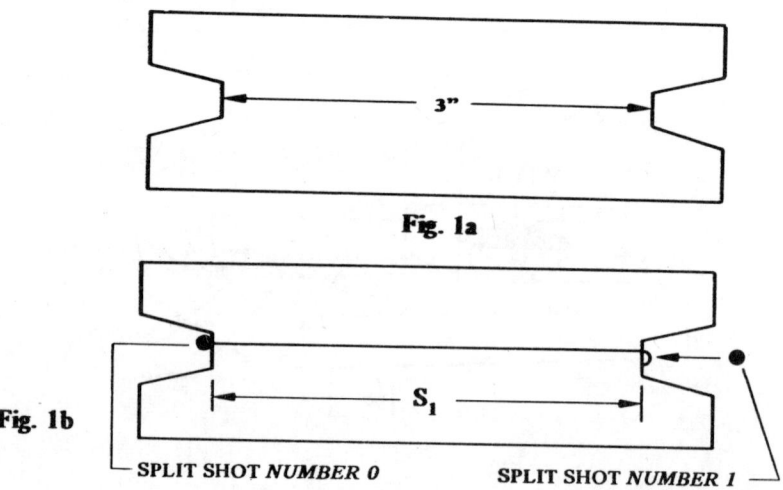

3"

Fig. 1a

S_1

Fig. 1b

SPLIT SHOT *NUMBER 0*   SPLIT SHOT *NUMBER 1*

using an oscilloscope and microphone, and the test results may form the basis for an interesting student exercise.

Assembly Directions

After crimping a piece of split shot on the end of a string, place the string in the V of the cardboard and pull it up tight. Crimp a second piece of shot on the string at the opposite end of the cardboard "string winder," as shown in Fig. 1b.

Next, wind the string around the "string winder" for three more lengths and fasten another piece of split shot on the string. Since one length was already on the string winder, the total length of the wound string is now $4S_1$. S_1 is the distance from the shot on the bottom of the string to the first piece of shot up the string.

Continue winding the string about the cardboard and fastening the shot according to the following intervals:

S_0 = the end of the string where shot *number zero* is fastened.
S_1 = the distance of shot number one (1) from the end of the string.
$S_2 = 2^2 S_1 = 4S_1$ measured from S_0.
$S_3 = 3^2 S_1 = 9S_1$ measured from S_0.
.
.
.
$S_n = n^2 S_1$.

Demonstration

By holding the completed string vertically with S_0 just touching the floor, the pieces of shot will be spatially arranged in such a way that, when dropped, the clicking sound of the split shot hitting the floor will be equally spaced *in time*. The sound can be easily amplified by dropping the weight on an inverted wastepaper can.

NOTE: This demonstration is analogous to the "dripping faucet" problems that are found in most physics texts.

Testing the Device

The apparatus may then be given a more sophisticated test with an oscilloscope and a microphone. Place the microphone on the floor with a metal plate over it, and drop the weights "onto the microphone." Apply the output signal of the microphone to the vertical amplifier of the oscilloscope so that the horizontal sweep will be triggered externally from the first impluse—S_1 hitting the metal plate.

Properly adjust the time base for the horizontal sweep so that all of the clicks can be recorded. Then, either by means of a storage scope or by photographing the trace on the scope, measure to see whether the peaks are equally separated.

> NOTE: This is a time separation and not a spatial separation, even though the distance between the peaks on the face of the scope is measured by a ruler!

Fig. 2 shows a typical trace where the time base was 50 msec/cm and S_1 was 3 inches.

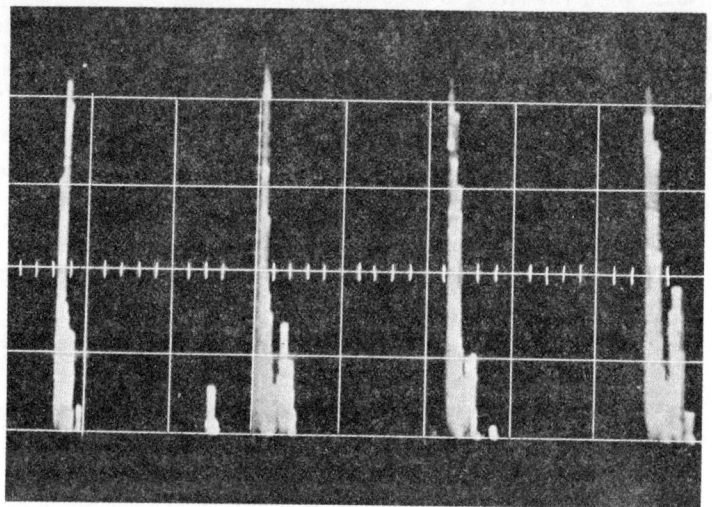

Fig. 2

Student Exercise

With the measurement data, students can complete the simple chart shown in Fig. 3. The numbers to be filled in by the students in column 1 represent the distance that the freely falling object traveled during the corresponding interval of time. The $\triangle S_4$, for example, is the distance traveled by the freely falling object during the fourth interval of time. By simply dividing $\triangle S_4$ by the time interval, one determines the average velocity of the particle during that interval.

Thus $\triangle S_4 / \triangle t = V_{ave}$ during the fourth time interval.

The numbers in column 2 represent the difference in the difference of the distance between intervals. Or more simply, if the differences from column 1 have been divided by the time interval, column 2 represents the average change in the velocity from interval to interval. And then the acceleration is easily calculated.

Thus
$$\frac{\triangle S_4 / \triangle t - \triangle S_3 / \triangle t}{\triangle t} = a.$$

A typical example, using S_1 = 3 inches and therefore t = 1/8 second, would yield the following:

$\triangle S_4$ = 1.75 ft. $\triangle S_3$ = 1.25 ft

$$\cfrac{\cfrac{1.75 \text{ ft}}{1/8 \text{ sec}} - \cfrac{1.25 \text{ ft}}{1/8 \text{ sec}}}{1/8 \text{ sec}} = 32 \text{ ft/sec/sec.}$$

By making S_1 = 3 inches, S_7 is 12.25 ft., which can be easily handled.

FOLLOW-UP: More ambitious students may want to have lengths of string long enough to drop down a stairwell or off the roof of a building. But take care, for the kinematics of a freely falling student could be disastrous!

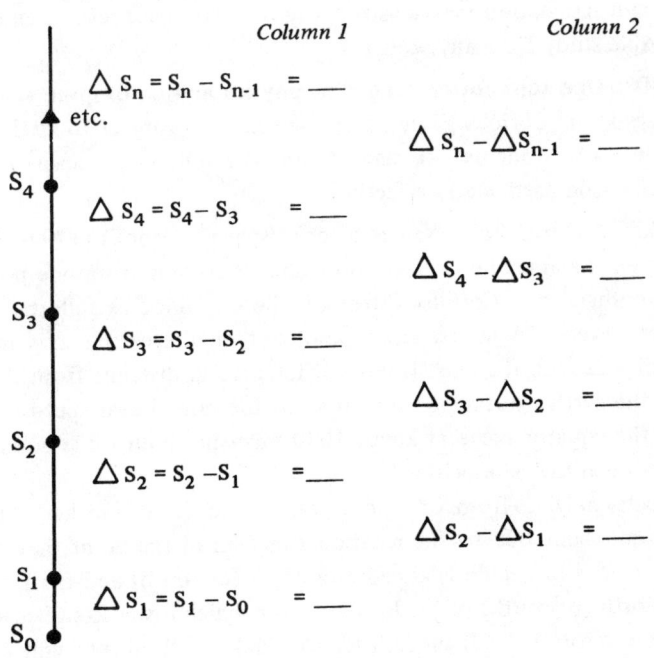

	Column 1	Column 2
	$\triangle S_n = S_n - S_{n-1}$ = __	
	etc.	$\triangle S_n - \triangle S_{n-1}$ = __
S_4	$\triangle S_4 = S_4 - S_3$ = __	
		$\triangle S_4 - \triangle S_3$ = __
S_3	$\triangle S_3 = S_3 - S_2$ = __	
		$\triangle S_3 - \triangle S_2$ = __
S_2	$\triangle S_2 = S_2 - S_1$ = __	
		$\triangle S_2 - \triangle S_1$ = __
S_1	$\triangle S_1 = S_1 - S_0$ = __	
S_0		

Fig. 3

THE CORIOLIS EFFECT—
AN INDUCTIVE APPROACH

by Harrie E. Caldwell

Wilkes College, Wilkes-Barre, Pennsylvania

The universe and earth's relationship to celestial bodies have always attracted men. However, in this era of rocketry and space travel, astronomy is more than simply fascinating; it is relevant to our lives. A combination of relevance and fascination makes astronomy or earth-space science an interesting and enjoyable study for many students.

NOTE: One topic often taught in physics and earth-space science classes is the Coriolis Effect. In presenting this subject to sixth and ninth grade students, we have found the following transparency-based lesson particularly effective.

Background materials: Named after Gaspar G. Coriolis (1792-1843), who first analyzed the apparent deflection of objects moving from one point on the earth to another,[1] the Coriolis Effect can be explained as follows: The earth rotates 360° every 24 hours. Each point not on the earth's axis moves in a circular point around the axis. Due to differences in distance from the axis, all points on the earth's surface do not move at the same linear speed or velocity. Points on the equator move at about 1040 miles per hour while points at other latitudes move at lower velocities.[2]

Objects moving from one point on the earth to another thus have a velocity component due to the rotational motion of the point they leave. For example, when a projectile is aimed due north (or south) and fired, it will not land due north (or south) of the location from which it was fired because of the velocity due to rotation.[3] If the latitude to which it is fired is moving eastwardly

---

[1] Earth Science Curriculum Project, *Investigating the Earth*, American Geological Institute, pp. 3-4.

[2] One may approximate the linear velocity V_L of the surface at any latitude using the formula: V_L V_O cos L (V_O is the linear velocity of a point on the equator and L is the latitude measured in degrees).

[3] Unless it lands in a corresponding latitude in the other hemisphere. (A projectile aimed due north and fired from $X^0 S$ latitude to $X^0 N$ latitude will land on the same meridian it left.)

at a faster rate than the latitude from which it left, the projectile will land west of the point at which it was aimed. If the latitude to which it is fired is moving slower than its point of origin, the projectile will land east of the point to which it was fired.

ILLUSTRATION: Fig. 1 shows paths of projectiles moving north or south above the earth. The projectiles were originally aimed due north (or due south). As illustrated, projectiles do not move in straight lines relative to the earth.

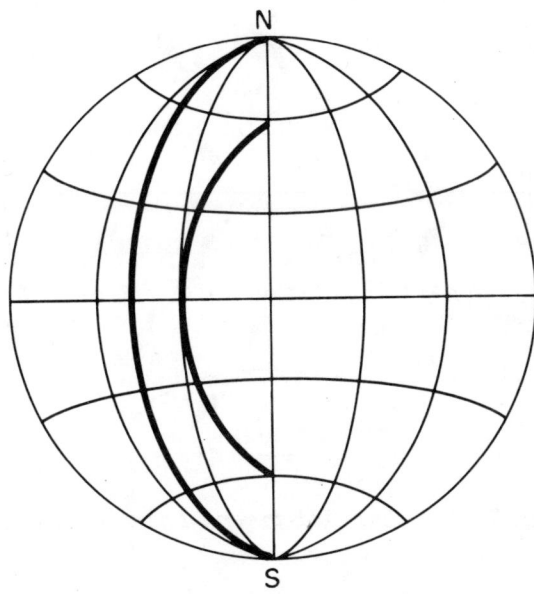

Fig. 1: *The heavy lines represent paths, relative to the earth, the projectiles aimed due north (or south) and fired would follow.*

This nonlinear motion of objects relative to a rotating coordinate system is often attributed to an imaginary force called "Coriolis Force." Since no real force is acting on objects, it is more appropriate to use the phrase "Coriolis Effect."

Examples: Following are several examples of the Coriolis Effect which are observable in our real world:

(1) Projectiles moving in a direction other than due east or due west appear to be deflected; those moving away from the equator seem to be deflected eastward and those moving toward the equator seem to be deflected westward.

(2)   Prevailing winds (Fig. 2) in various sectors occur because the earth's surface is heated unevenly. At the equator, warmed air rises and flows toward the poles. This air, cooled at higher altitudes, settles eastward near latitude 30° and flows north and south over the earth's surface. However, these air masses do not flow due north or due south.

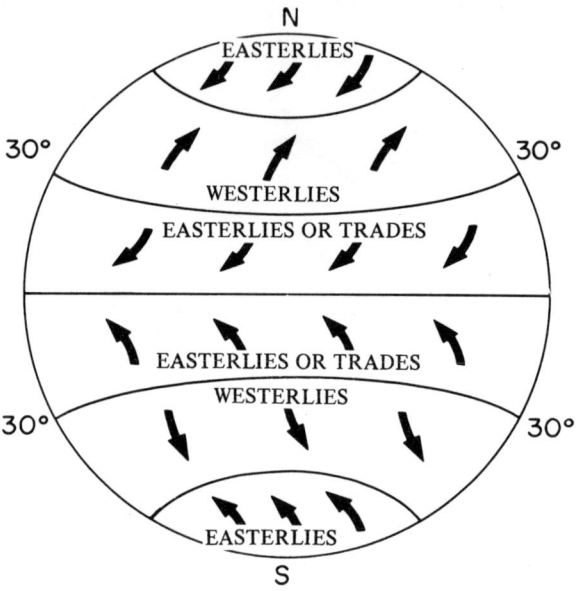

Fig. 2: *Prevailing winds in various parts of the earth.*

Those masses moving toward the equator appear to come from an easterly direction (they are deflected to the west) and are called "easterlies" or "trades." Those moving away from the equator appear to come from a westerly direction (they are deflected to the east) and are called "westerlies." Also, cold polar air masses moving toward the equator appear to be deflected westward and are called "easterlies."

(3)   Cyclonic winds (around low pressure areas) move counterclockwise in the northern hemisphere and clockwise in the southern hemisphere. Anticyclonic winds (around high pressure areas) move in the opposite direction.

NOTE: Another expression of the Coriolis Effect, sometimes called Ferrel's Law, is: Projectiles or winds are deflected to the right in the northern hemisphere and to the left in the southern hemisphere.

Lesson: The Coriolis Effect

This lesson is centered around a set of nine simple transparencies for use on an overhead projector. Following is a description of each transparency and suggestions for its effective use.

Transparency 1 (Fig. 1): Showing of Transparency 1, which illustrates the paths of projectiles moving north or south above the earth, may be accompanied by two statements:

1. The sun moves across the sky every day.
2. A pendulum is set in motion. Although gravity is the only force to act on the pendulum, the plane of the pendulum's swing is observed to turn relative to the earth.

These two statements provide data from which students may infer rotation of the earth.

The teacher may wish to have students discuss implications of the two statements immediately. Or he may want to give students the opportunity to relate this data to the problems presented in later transparencies before any discussion is held. The latter approach seems appropriate for average or above average students, while the former approach might be better for slower students.

Transparency 2 (Fig. 2): Transparency 2 shows the direction of prevailing winds in various sectors of the earth. The direction of prevailing winds is explained by the Coriolis Effect. This diagram should be shown again later when students have generalized the concept Coriolis Effect to give them an opportunity to apply their generalization. It should not be discussed before students have generalized the concept.

Transparencies 3-9 (Figs. 3-9): Each of the remaining transparencies actually consists of two transparencies connected so that one side (side B) can be flipped on top of the other (side A). Side A in each case poses a problem. It contains a diagram of the earth, a rocket, and three points labeled A, B, and C. The rocket is aimed due east, due north, etc., and fired. Point B is always in line with the direction the rocket is aimed; points A and C are located on either side of point B. A caption reads: "a. A rocket is aimed due . . . and fired.   b. Will the rocket land at A, B, or C?"

When side B is placed on top of side A, students are provided with the correct answer. Side B contains a black spot to cover the rocket on side A and a diagram of a rocket which makes it appear that the rocket's nose is buried in the ground. The rocket on side B is positioned to coincide with the correct solution (A, B, or C) on side A.

The first time transparencies 3-9 are shown, students should not discuss them. Show the transparencies. When students have had sufficient opportunity to react to the solution (side B) on one transparency, show the problem (side A) with the next transparency. Continue this procedure until students have reacted

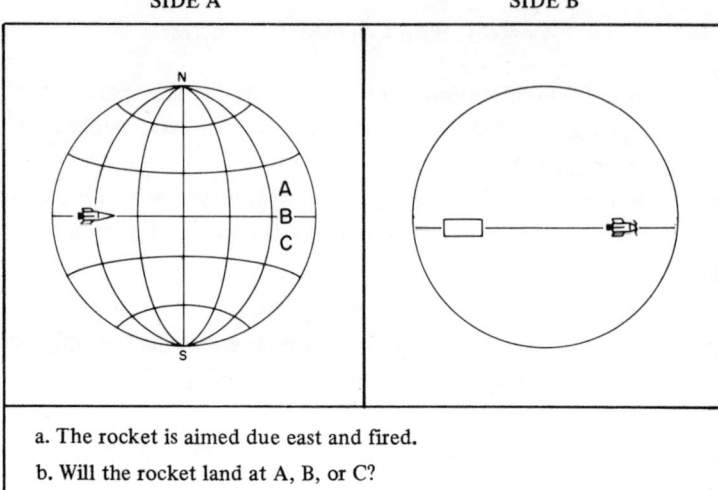

a. The rocket is aimed due east and fired.

b. Will the rocket land at A, B, or C?

Fig. 3: **(Transparency 3)** *The rocket is located on the equator and aimed due east. Points A, B, and C are located in the east. The rocket will land at point B, which is due east relative to the point from which the rocket was fired.*

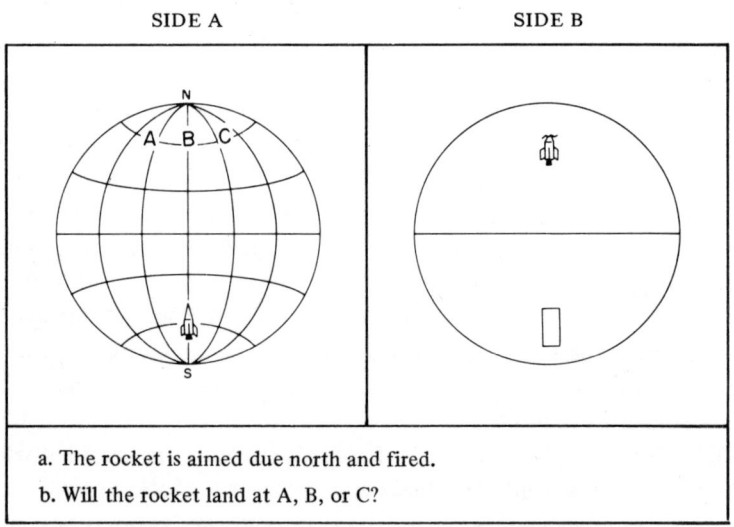

a. The rocket is aimed due north and fired.

b. Will the rocket land at A, B, or C?

Fig. 4: **(Transparency 4)** *The rocket is located near the South Pole and aimed due north. Points A, B, and C are located near the North Pole. The rocket will land at point B because the eastward velocity (due to rotation) of points A, B, and C is the same as the eastward velocity (due to rotation) of the rocket.*

SIDE A   SIDE B

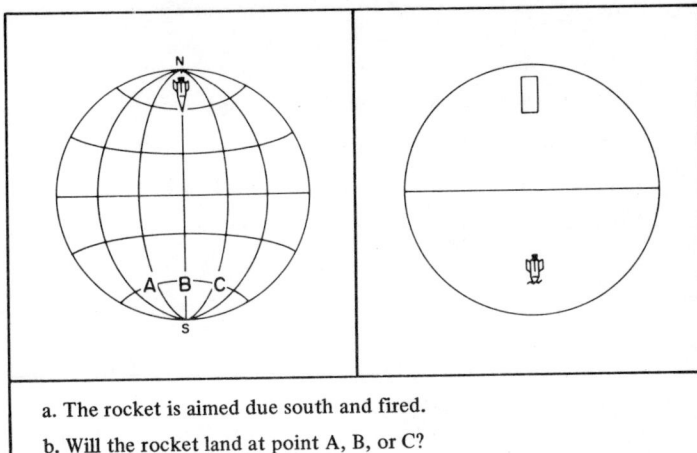

a. The rocket is aimed due south and fired.

b. Will the rocket land at point A, B, or C?

Fig. 5:   (Transparency 5) *The rocket is located near
the North Pole and aimed due south. Points A,
B, and C are located near the South Pole. The
rocket will land at point B because the eastward velocity (due to rotation) of points A, B,
and C and the eastward velocity (due to rotation) of the rocket are equal.*

SIDE A   SIDE B

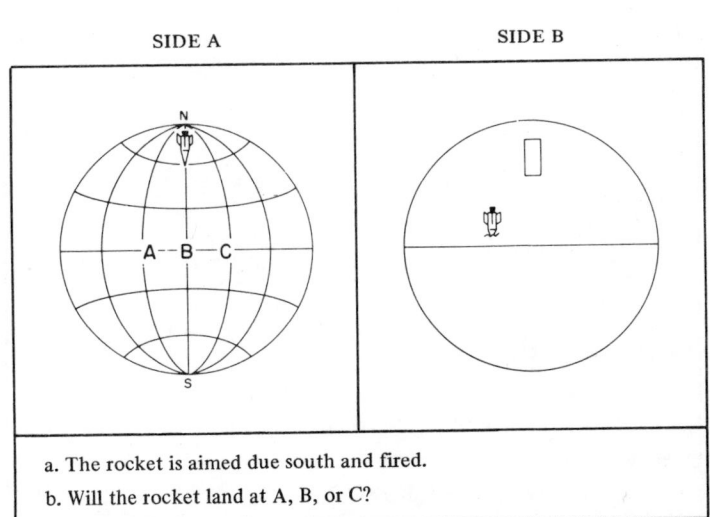

a. The rocket is aimed due south and fired.

b. Will the rocket land at A, B, or C?

Fig. 6:   (Transparency 6) *The rocket is located near the
North Pole and aimed due south. Points A, B,
and C are located on the equator. The rocket
will land at point A because the eastward
velocity of points A, B, and C is greater than
the eastward velocity of the rocket.*

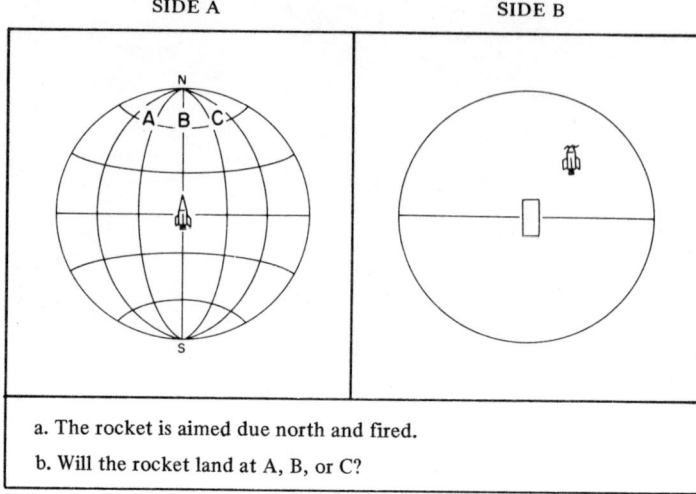

Fig. 7: **(Transparency 7)** *The rocket is located near the equator and aimed due north. Points A, B, and C are located near the North Pole. The rocket will land at point C because the eastward velocity of points A, B, and C is less than the eastward velocity of the rocket.*

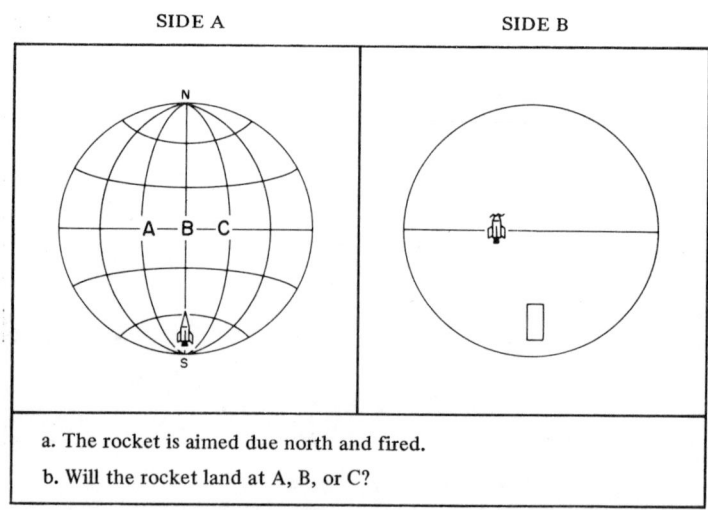

Fig. 8: **(Transparency 8)** *The rocket is located near the South Pole and aimed due north. Points A, B, and C are located on the equator. The rocket will land at point A because the eastward velocity of points A, B, and C is greater than the eastward velocity of the rocket.*

SIDE A   SIDE B

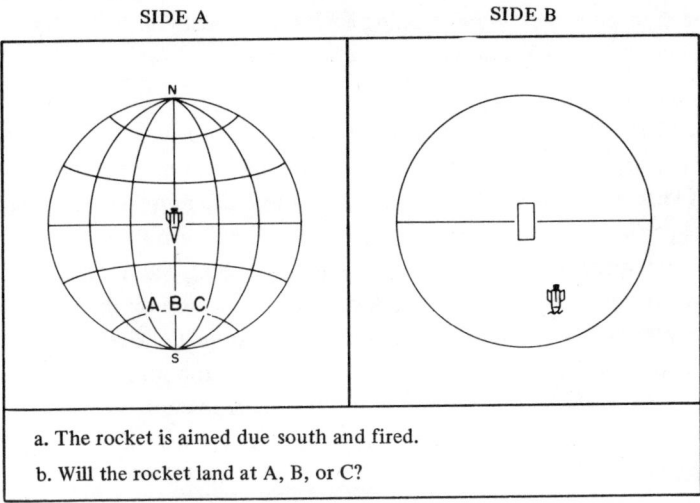

a. The rocket is aimed due south and fired.

b. Will the rocket land at A, B, or C?

Fig. 9: (**Transparency 9**) *The rocket is located near the equator and aimed due south. Points A, B, and C are located near the South Pole. The rocket will land at point C because the eastward velocity of points A, B, and C is less than the eastward velocity of the rocket.*

to all of the problems and observed all of the solutions. Then let them choose transparencies they wish to see again.

> **NOTE: Teachers should be careful not to allow one student to solve a problem aloud before other students have had an opportunity to solve it for themselves.**

Using the Inductive Approach

Learning: It is not suggested that each student will be able to verbalize the Coriolis Effect from these transparencies. However, when a student is able to predict answers consistently, he probably has an internalized concept, at least at a sub-verbal level. When this student then hears a verbal statement of the Coriolis Effect, he should be able to associate the verbal statement with his sub-verbal concept.

It would be easy to find out which students are able to verbalize statements of the concept. Before discussing data, have students write a statement(s) explaining how they solved the problems or how they would explain the answers to someone else.

DISCUSSION: The final step is to discuss the problems and conclusions with the class. This will probably entail going back over the transparencies and discussing each transparency individually and, also, relating each transparency to the others.

Variations: If an overhead projector is unavailable, the diagrams might be reproduced on cards or papers, perhaps in booklet form. Students could work individually at their own rates, or in small groups with other students of similar ability. (They might also take materials home and attempt to teach their parents the concept. Teaching the concept may help them to learn it.)

To get students to think about this phenomenon, the transparencies might be shown without discussion several days prior to a lesson on the Coriolis Effect. For example, the lesson could be scheduled for a Monday. The overlays could be shown during the last ten minutes of the period on the preceding Friday. Some students will undoubtedly forget about the phenomenon as soon as the bell rings, but others may be intrigued. The latter may even work on the puzzle over the weekend.

8

PERIODIC MOTION AND SPACE PHYSICS

SPACE FLIGHT EXERCISES
FOR HIGH SCHOOL STUDENTS

by Harry Lobel

Formerly Thomas Jefferson High School, Council Bluffs, Iowa

The launching of the space age has created a new generation of science drop-outs—students who are interested in space problems but handicapped in exploring them by the incomprehensible symbols involved. To overcome this difficulty, we developed a series of exercises dealing with the flights of satellites and the motions of planets, using symbols which are simple but challenging. Instead of employing miles per hour, for example, they use miles per second. Thus the student is working with an escape velocity of 7 miles per second rather than 25,200 miles per hour. Also the student is cautioned against trying for too great a degree of accuracy, and in no case are more than three significant figures employed.

> USE: These space flight exercises may be thought of as enrichment for courses in geometry, senior science, and physics. To avoid further burdening of already overloaded courses, the teacher can work and quiz students in groups or teams consisting of a Captain, Pilot, Co-pilot, and Navigator.

Materials needed to perform the various activities described are simple. Every student should be given a polarized work sheet with concentric circles at 1 centimeter intervals and 12 radial segments of 30 degrees each. These can be easily reproduced on a ditto machine. The drawing of an ellipse calls for only a compass. The exercises on velocities and periods are solved by four-place logarithms or a slide rule.

1. **Johannes Kepler's discoveries**: Johannes Kepler (1571-1630) believed that all planets, including the Earth, revolved about the Sun (Figs. 1 and 2). However, prior to his time it was not possible to explain: (a) why the planets seemed to have retrograde motion; (b) why the velocities of some planets varied in different parts of their orbits. Through Kepler's discoveries, we know today that both of these problems are accounted for by the fact that the closer it is to the Sun, the faster a planet revolves.

Retrograde motion is due to the fact that the velocity of the planet Mars, for example, is less than the velocity of the Earth (Fig. 3). Mars is seen against the backdrop of the stars. As the Earth overtakes and passes Mars, the latter

Fig. 1: Earth-Centered System

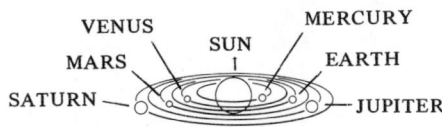

Fig. 2: Sun-Centered System

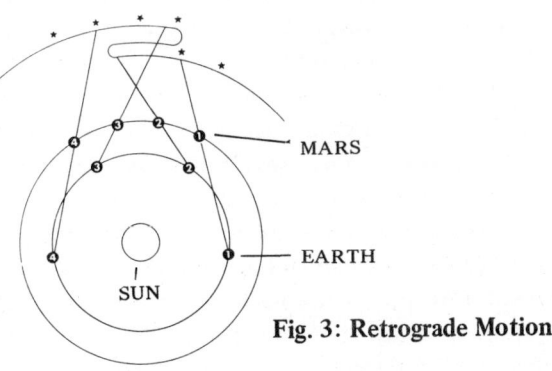

Fig. 3: Retrograde Motion

seems to stop, reverse direction, stop again, and then move forward. Retrograde motion is thus an illusion caused by parallax as the slower planets are overtaken by the faster-moving Earth.

The solution to the second problem was more difficult for Kepler. Why do the velocities of Mercury and Mars, for example, vary in different parts of their orbits? Kepler was handicapped by an ancient belief that the orbits of the planets were perfect circles. If the paths were perfect circles, how could the velocities vary?

2. **The conical pendulum:** The nature of this problem can be demonstrated by a conical pendulum in which the bob is made to follow a circular path (Fig. 4). To ensure that the path of the bob is circular, the pendulum cord is passed through the stem of a funnel, the lip of the funnel serving as a guide. This particular type of pendulum shows that when the path of the bob is circular, the velocity is constant in every part of the orbit.

Kepler was astute enough in mathematics and mechanics to realize that the orbits of some planets, especially Mercury and Mars, were other than circular. What type of geometrical path other than a circle could the planets follow?

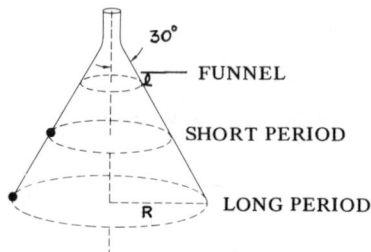

Fig. 4: The Conical Pendulum

3. **Keplerian motion**: After trying many geometrical patterns, Kepler discovered that the orbits of planets could be elliptical as well as circular. According to this discovery, as the distance from the planet to the Sun increased, the velocity decreased. However, there was another complicafion which made the solution of this problem more difficult. In Kepler's day, just as today, the geometry books described the ellipse as being symmetrical about its center. The significance of this is that if the Sun were at the center of the elliptical paths of the planets, there would be an increase in velocity twice each time they revolved about the Sun, instead of just once.

Thus Kepler had to visualize an elliptical path with the Sun at a focal point rather than at the center, as shown in Fig. 5. In this illustration, the Sun is at one of the focal points. When a planet is closest to the Sun, at the perihelion, its velocity is greatest. As it travels from the perihelion toward the farthest point, its velocity diminishes. At the aphelion, the planet's velocity is least.

> **NOTE**: This change in velocity in elliptical motion is known as Keplerian motion, and is applicable to the planets, satellites, and space ships. Keplerian motion is, in fact, characteristic of the motion of any object orbiting in an ellipse about a central body.

4. **Characteristics of the ellipse**: Our problem—just as it was Kepler's problem—is to draw an ellipse with the Sun at a focal point, and to demonstrate

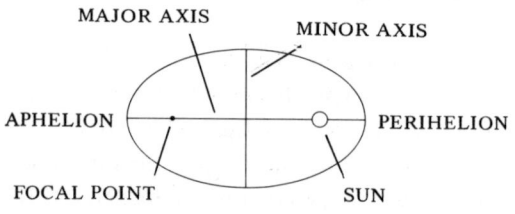

Fig. 5: Elliptical Path

a change in velocity which varies inversely with the distance from the Sun.

(a)   The largest diameter of an ellipse is called the *major axis;* the shortest diameter is the *minor axis.*

(b)   The end of the major axis which is closest to the Sun is the perihelion, *peri-* meaning *around.* The end of the major axis farthest from the Sun is the aphelion, *ap-* meaning *away from.*

(c)   The force of gravity is greatest at the perihelion and least at the aphelion. Centrifugal force and, therefore, the velocity of the planet is greatest when nearest the Sun and least when farthest from the Sun.

The ellipse has a characteristic which facilitates an accurate geometrical construction: The sum of the focal distances from any point on the ellipse is constant; this sum is always equal to the major axis. For instance, in Fig. 6 the sum of XD plus DS equals XE plus ES, equals XF plus FS. And these sums are each equal to the major axis, AP. This characteristic can be employed to construct the trajectory of a space ship traveling from the orbit of one planet to the orbit of a second planet.

5. **Construction of the ellipse:** Assume that we are to trace the trajectory of a space ship from a perihelion of 100,000,000 miles to an aphelion of 500,000,000 miles. This would correspond to orbits which are slightly larger than orbits of the Earth and of Jupiter. For convenience, we may assume that the distances between the circular orbits are 50,000,000 miles.

> **NOTE: This information is all that is necessary for the construction of a reasonably accurate trajectory. It allows us to determine the number of miles traveled by a space ship as well as the distance of Jupiter from the point of rendezvous at the time of blast-off.**

The location of the Sun and of the perihelion and aphelion are given by the statement of the problem. The Sun corresponds to the origin or center of the polar chart. Two spaces from the sun, at the end of the major axis, is the perihelion. Ten spaces from the Sun, at the far end of the major axis, is the aphelion. An imaginary focal point is on the major axis, two spaces from the aphelion.

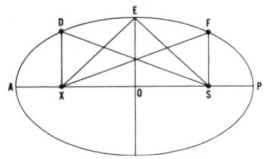

Fig. 6: Characteristics of the Ellipse

For practical purposes, the determination of only eight points is all that is necessary for a reasonably accurate trajectory. The ends of the major axis are given by the statement of the problem. The ends of the minor axis, EH, are found by determining the midpoint of the major axis, and measuring this distance from the focal points, X and S, as seen in Fig. 7. Four other points—D,

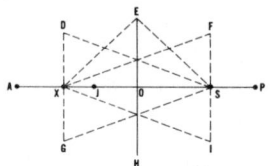

Fig. 7: Construction of the Ellipse

F, G, and I—are determined in a similar manner. Instead of measuring to the midpoint of the major axis, however, some other convenient point, J, is chosen between one of the focal points and the midpoint, O.

(a)   D is found by measuring the distance of AJ from focal point X, and PJ from focal point S.

(b)   F is determined by measuring AJ from focal point S, and PJ from focal point X.

This determines eight points: the perihelion, the aphelion, the ends of the minor axis, and D, F, G, and I. The ellipse is sketched free-hand using these points as guides.

> NOTE: While there is no limit to the number of guide points which can be determined, in my classes I tried to discourage more than an approximate degree of accuracy. Since this is only an approximation of the trajectory of a space flight, it seems best to avoid instilling a sense of false accuracy in the minds of students.

6. Periods of flight: With a reasonably accurate ellipse, the distances traveled by a space ship in equal periods of time can be visualized. Tentatively, the points along the lower half of the ellipse, where the radial lines cross the ellipse, are labeled in a counterclockwise direction as A', B', C', D', E', and F', as in Fig. 8. This indicates a large velocity at A'B', which is close to the aphelion, and a diminishing velocity at B'C', C'D', D'E', and E'F', as the space ship approaches the perihelion. Finally, at the perihelion, the distance F'P indicates a minimum velocity.

According to Kepler's second law, the velocity at the perihelion should be greatest and that at the aphelion should be least. Therefore, to correct the diagram, distance A'B' is transferred diagonally to PB, B'C' is transferred to BC, C'D' is transferred to CD, and so on. Now the distances covered in equal periods of time are greatest at the perihelion and least at the aphelion.

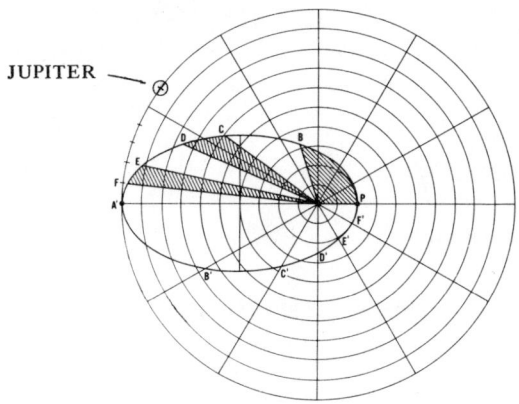

Fig. 8: Trajectory from Earth to Jupiter

JUPITER: To determine the location of planet Jupiter at the time of blast-off from the Earth's orbit, the distance FA' is measured six times along the orbit of Jupiter.

7. Earth satellites: A satellite orbiting about the earth is subject to the same laws as a planet revolving about the Sun. Ordinarily, the satellite follows either a circular or an elliptical orbit. In the event the satellite achieves an escape velocity, it follows a parabolic trajectory as in Fig. 9. The Earth is at the focal point of the satellite's normal elliptical orbit. The end of the major axis nearest the Earth is the *perigee;* the end of the major axis farthest from the Earth is the *apogee.* A convenient value for distances between the concentric circles in Fig. 10 is 4,000 miles, half the approximate diameter of the Earth.

Conventionally, the altitude of satellites is measured from the Earth's surface. However, in determining the trajectory, the altitude above the center of the Earth must be calculated. For simplicity in symbols, in the accompanying table (Table 1) the altitude above the Earth's center is given in terms of radii.

Table 1

Symbols for Altitude above Earth

Altitude above Surface	*Altitude above Center*	*Altitude in Radii*
0 miles	*4,000 miles*	*1*
4,000 miles	*8,000 miles*	*2*
8,000 miles	*12,000 miles*	*3*
12,000 miles	*16,000 miles*	*4*

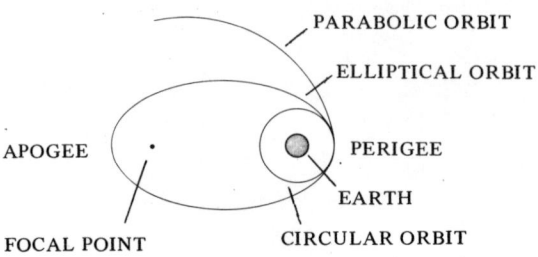

Fig. 9: Satellite Orbits

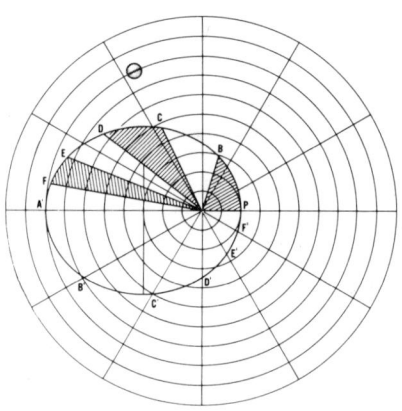

Fig. 10: Trajectory of a Space Ship

As an example of a space flight, a capsule in orbit 8,000 miles above the surface of the Earth is approximately 12,000 miles or three radii from the center. This capsule is to rendezvous with a space ship 28,000 miles above the surface of the Earth, and at an altitude of 32,000 miles or eight radii from the center. In this case:

(a) The major axis extends from the perigee, three spaces from the center of the Earth, to the apogee, eight spaces to the left of the center.

(b) The imaginary focal point is 12,000 miles or three spaces to the right of the apogee.

(c) The center of the major axis is 10,000 miles or two and a half spaces from the Earth's center.

As in a flight between planets, first the points of the minor axis are determined, then the four other points on the ellipse. And just as in the case of motion between planets, the intersection between the radial lines and the lower half of the ellipse (points A', B', C', D', E', and F') are transferred diagonally from the lower to the upper half of the ellipse. This indicates the distances traveled in each of six periods of time. The location of the space ship at the time of blast-off is six times the distance FA'.

8. **Keplerian motion and the conical pendulum**: The analog of velocities of planets and space ships can be demonstrated by means of the conical pendulum.

When the angle of the conical pendulum is guided by a funnel (60°), the period of the conical pendulum and that of a simple pendulum of the same length are identical. The period of the simple pendulum, T, equals $2\pi\sqrt{1/g}$; "1" representing the length and "g", gravity. In the case of the conical pendulum, our main interest is the analog of planetary and satellite motions. Therefore, our concern is in the relation between the period and the radius rather than the relation between period and length. Since r/1 equals sin 30°, then "1" equals r/sin 30°, and T equals $2\pi\sqrt{r/g \sin 30°}$.

Thus a conical pendulum of 10 inches length has a radius of 5 inches and a period of 1 second. A conical pendulum of 40 inches length has a radius of 20 inches and a period of 2 seconds. From these examples it can be seen that the smaller the radius, the shorter the period.

ALGEBRAIC EXERCISES: In developing algebraic exercises, described in the remainder of the article, simple symbols were sought. All exercises were designed for four-place logarithms or slide-rule solution, and in no case were more than three significant figures employed.

9. **The planetary orbits**: For determining the distances of the planets from the Sun, Bode's Law is used. This gives an approximate distance in Astronomical Units of 93,000,000 miles each.

Table 2

Planetary Distances

Planet	Bode's Law	Approximate Distance	Actual Distance
Mercury	0 x 0.3 + 0.4	0.4	0.39
Venus	1 x 0.3 + 0.4	0.7	0.72
Earth	2 x 0.3 + 0.4	1.0	1.0
Mars	4 x 0.3 + 0.4	1.6	1.52
Asteroids	8 x 0.3 + 0.4	2.8	2.8
Jupiter	16 x 0.3 + 0.4	5.2	5.2
Saturn	32 x 0.3 + 0.4	10.0	9.55
Uranus	64 x 0.3 + 0.4	19.6	19.2

Neptune	30
Pluto	40

10. The velocity of planets: The velocities of the planets are determined as a ratio of the velocity of the Earth. Their distances are measured in Astronomical Units, either according to Bode's Law or to the actual A.U. The velocity of the Earth is approximately 18.5 miles per second.

$$Vp/Ve = \sqrt{Re/Rp} \qquad (1)$$
$$Vp/18.5 = \sqrt{1/Rp} \qquad (2)$$
$$Vp = 18.5/\sqrt{Rp} \qquad (3)$$

Examples:

Mercury	$18.5/\sqrt{0.4} = 18.5/0.63 = 29.3$ mps
Venus	$18.5/\sqrt{0.7} = 18.5/0.83 = 22.2$ mps
Earth	$18.5/\sqrt{1.0} = 18.5/1.0 = 18.5$ mps
Mars	$18.5/\sqrt{1.6} = 18.5/1.26 = 14.6$ mps
Asteroids	$18.5/\sqrt{2.8} = 18.5/1.67 = 11.0$ mps

11. The periods of planets: The periods of the planets are determined as ratios of the Earth's period.

$$(Tp)^2 / (Te)^2 = (Rp)^3 / (Re)^3 \qquad (4)$$
$$Tp/Te = Rp\sqrt{Rp} / Re\sqrt{Re} \qquad (5)$$
$$Tp = Rp\sqrt{Rp} \qquad (6)$$

As examples of periods:

Mercury	$0.39\sqrt{0.39} = 0.24$ years $= 87.6$ days
Venus	$0.72\sqrt{0.72} = 0.60$ years $= 219$ days
Mars	$1.52\sqrt{1.52} = 1.82$ years
Asteroids	$2.8\sqrt{2.8} = 4.7$ years

12. The velocities of satellites: Velocities of satellites orbiting about the Earth are determined as ratios of the approximate velocity of a satellite at the surface of the Earth.

$$Ve \simeq 5 \text{ mps} \qquad (7)$$
$$Vs/Ve = \sqrt{Re/Rs} \qquad (8)$$
$$Vs/5 = 1/\sqrt{Rs} \qquad (9)$$
$$Vs = 5/\sqrt{Rs} \qquad (10)$$

Examples of satellite velocities:

Distance from Center of Earth		Approximate Velocity
4,000	mi	$5/\sqrt{1} = 5.0$ mps
8,000	mi	$5/\sqrt{2} = 3.5$ mps

12,000 mi	$5/\sqrt{3} = 3.0$ mps
16,000 mi	$5/\sqrt{4} = 2.5$ mps
100,000 mi	$5/\sqrt{25} = 1.0$ mps

13. The periods of satellites: The periods of satellites are determined as ratios of the approximate period of a satellite at the surface of the Earth.

$$(Ts)^2 / (Te)^2 = (Rs)^3 / (Re)^3 \qquad (11)$$
$$Ts/Te = Rs\sqrt{Rs}/Re\sqrt{Re} \qquad (12)$$
$$Ts/1.5 = Rs\sqrt{Rs} / \sqrt{1} \qquad (13)$$
$$Ts = Rs\sqrt{Rs} \times 1.5 \text{ hours} \qquad (14)$$

Examples of satellite periods:

Distance from Center of Earth	Periods of Satellites
4,000 mi	$1\sqrt{1} \times 1.5 = 1.5$ hours
8,000 mi	$2\sqrt{2} \times 1.5 = 4.2$ hours
12,000 mi	$3\sqrt{3} \times 1.5 = 7.65$ hours
16,000 mi	$4\sqrt{4} \times 1.5 = 12.0$ hours
26,000 mi	$6.5\sqrt{6.5} \times 1.5 = 24.7$ hours

14. The Hohmann transfer velocity: Another exercise which leads to an understanding of Kepler's second law is the flight from the orbit of the Earth to the orbit of a second planet. When this flight sweeps out 180° and is from a perihelion to an aphelion, it is known as a Hohmann flight or a Hohmann transfer, named after Walter Hohmann, who developed the mathematics in 1925.

The Hohmann transfer velocity is approximately proportional to the inverse of the perihelion and aphelion distances. For example, the transfer velocity for a flight from the Earth's orbit to the orbit of Mars is:

$$Vtem = (18.5/\sqrt{a}) \times (\sqrt{AP}) . \qquad (15)$$

These symbols are as follows:

V indicates velocity
t " transfer
e " Earth
m " Mars
a " ½ major axis
A " aphelion
P " perihelion

$$Vtem = 18.5/\sqrt{(1)(1.6)/2} \times \sqrt{A/P} \qquad (16)$$
$$Vtem = 18.5/\sqrt{1.3} \times \sqrt{1.6} \qquad (17)$$
$$Vtem = 20 \text{ mps.} \qquad (18)$$

The significance of this calculation is that a space ship traveling to the Earth's orbit would have to increase its velocity from 18.5 mps to 20 mps to reach the orbit of Mars.

15. The transfer velocity to Venus: The transfer velocity to Venus would be:

$$Vtev = 18.5/\sqrt{a} \times \sqrt{A/P} . \qquad (19)$$

The new symbols are as follows:
     e indicates Earth
     v   "   Venus

$$Vtev = 18.5/\sqrt{(1 + 0.7)/2} \times \sqrt{0.7/1} \quad (20)$$
$$Vtev = 18.5/\sqrt{0.85} \times \sqrt{0.7} \qquad (21)$$
$$Vtev = 16.6 \text{ mps.} \qquad (22)$$

In this case, the calculations indicate that a space ship traveling in the Earth's orbit at 18.5 mps would have to fire its retrorockets to decrease its velocity from 18.5 mps to 16.6 mps to reach the orbit of Venus.

16. The transfer velocity of earth satellites: In transferring from one orbit to another about the Earth, the formula is:

$$Vt = 5/\sqrt{a} \times \sqrt{A/P}. \qquad (23)$$

To transfer from an orbit of 12,000 miles (three radii) to an orbit of 20,000 miles (five radii):

$$Vt35 = 5/\sqrt{(3+5)/2} \times \sqrt{5/3} \qquad (24)$$

V indicates velocity
t   "   transfer
3   "   three radii
5   "   five radii

$$Vt35 = 3.2 \text{ mps.} \qquad (25)$$

The converse, to transfer from the fifth radius to the third:
$$Vt53 = 5/\sqrt{(5+3)/2} \times \sqrt{3/5} \qquad (26)$$
$$Vt53 = 1.9 \text{ mps.} \qquad (27)$$

The significance of the calculations here is that to transfer from an orbit equal to three radii to an orbit equal to five radii, the velocity must be increased to 3.2 mps; to transfer from the fifth radius to the third radius, the velocity must be reduced to 1.9 mps.

As another example, to transfer from six radii to four:

$$Vt64 = 5/\sqrt{5} \times \sqrt{4/6} = 1.8 \text{ mps} \tag{28}$$
$$Vt46 = 5/\sqrt{5} \times \sqrt{6/4} = 2.7 \text{ mps.} \tag{29}$$

17. Kepler's second law: According to Kepler's second law, a space flight in an elliptical orbit sweeps out equal areas in equal periods of time. The velocity at the perigee times the radius of the perigee is equal to the velocity at the apogee times the radius of the apogee.

$$VpRp = V_A R_A \tag{30}$$

In the velocities of (25) and (27) above:

$$Vt35 = 3.2 \text{ mps} \tag{25}$$
$$Vt53 = 1.9 \text{ mps} \tag{27}$$
$$3.2 \times 3 = 1.9 \times 5 \tag{31}$$
$$9.6 = 9.5 \ . \tag{32}$$

In the velocities of (28) and (29) above:

$$Vt64 = 1.8 \text{ mps} \tag{28}$$
$$Vt46 = 2.7 \text{ mps} \tag{29}$$
$$1.8 \times 6 = 2.7 \times 4 \tag{33}$$
$$10.8 = 10.8. \tag{34}$$

DEVELOPMENT OF THE INVERSE-SQUARE LAW FOR ELLIPTICAL ORBITS

by David L. Speer

Science-Math Supervisor, Shaler Township Schools,
Glenshaw, Pennsylvania

In this exercise, students follow the steps necessary to prove graphically the inverse-square law for elliptical orbits. The method used is first demonstrated by Dr. A.V. Baez in his film, "Elliptic Orbits," but the direct involvement provided by the exercise is necessary to make it fully meaningful to students.

Preparation: Before undertaking the exercise, students should be thoroughly familiar with Kepler's laws concerning planetary orbits:

1. Each planet moves in an elliptical path with the sun at one focus.

2. The line joining the sun and planet sweeps out equal areas in equal times.
3. The ratio R^3/T^2 is the same for all planets (if this constant ratio is called K, we may say $R^3/T^2 = K$).

Students should know from their study of centripetal force that the centripetal force (F) on any object moving in a circular path may be found by the use of the equation $F = ma = m^4\pi^2 R/T^2$. T may be eliminated from the equation and the force expressed as a function of R and m alone by using Kepler's third law:

$$R^3/T^2 = K, \text{ or } T^2 = R^3/K. \text{By substituting } R^3/K \text{ for } T^2,$$
$$\text{we find } F = 4\pi^2 Km/R^2.$$

Thus the force is proportional to the mass revolving and inversely proportional to the square of its distance from the sun.

Presenting the problem: The teacher can explain that what has been found in the preceding calculations holds true for objects traveling around a circular path. However, we know from Kepler's first law that planetary orbits are not circular but elliptical. Satellites are constantly sampling the force field at different distances.

PROBLEM: Does the inverse-square relationship of force and distance hold true for elliptical orbits?

Student Exercise

Materials: To complete this activity, each student will require these materials:

1 sheet of 36" x 24"paper
2 thumbtacks
String
3 rulers, 2 taped together in a V-shape
Pencil
Small lead shot (or other suitable medium)

Procedure: Student directions are as follows:

(1) Take the large sheet of paper and construct an ellipse on it by placing two thumbtacks as foci and connecting them with a piece of string, as shown in Fig. 1. Keeping the string taut, draw an ellipse.

(2) Designate a length, R, and draw three lengths: R, 2R, and 3R (Fig. 2).

(3) Construct a tangent to the ellipse where each radius intersects the ellipse (Fig. 3).

QUERY: Kepler's second law states that equal areas are swept out in equal times. How can we relate this statement to our ellipse?

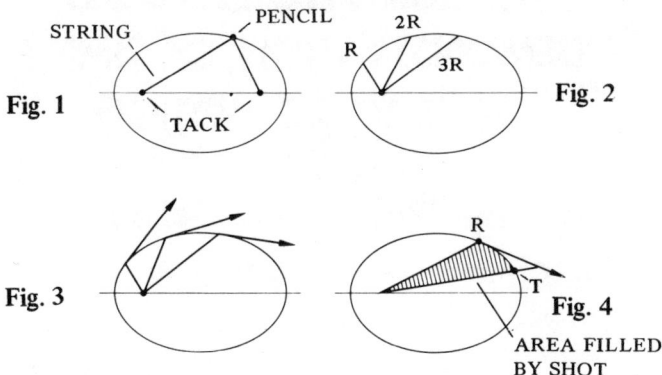

Fig. 1   Fig. 2

Fig. 3   Fig. 4

AREA FILLED
BY SHOT

(4) Choose an arbitrary time, T, a distance from the radius, R, on the ellipse, and mark both spots on the circumference. Draw a line from the focus through point T, and extend the line out until it intersects the tangent (Fig. 4).

(5) Now fill the area between line T and line R to the circumference of the ellipse with small lead shot, using the two rulers taped in a V-shape to contain the shot.

(6) Use the lead shot and the taped rulers to form two other areas equal to the first for 2R and 3R, and mark off T_2 and T_3 (in accordance with Kepler's second law). Continue the lines to the tangent as for R.

QUERY: What does the extension from T to the tangent represent?

(We know that if the object were to be released at the point where the radius intersects the ellipse, it would travel off along the line tangent to that point. Thus the line we have drawn from the tangent to point T represents the difference between the two paths, or the centripetal acceleration. Since F = ma and m is constant, we can use this centripetal acceleration to represent force on the object.)

(7) Measure the values of the acceleration for R, 2R, and 3R.

NOTE: We are trying to find the relationship between centripetal force and the radius.

(8) Compare the radii with the values obtained by measuring the acceleration. How do they compare numerically?

(9) Now compare by multiplying F x R^2. Does F x R^2 equal a constant? Does the inverse-square law also apply to elliptical orbits? Can you see why?

THE RUNNING SHADOW DEMONSTRATION OF SIMPLE HARMONIC MOTION

by Robert A. Schley

San Jacinto Community College, Gilman Hot Springs, California

An interesting phenomenon of mechanics, often slighted in descriptive physics courses, is simple harmonic motion. This type of motion is important because it occurs so frequently in nature. The word "simple" is intended to describe the geometry of the harmonic motion rather than the anticipated difficulty of understanding it.[1]

Demonstration

To convey the physical meaning of simple harmonic motion, in which the acceleration of a body is directly proportional to the body's displacement, and in the opposite direction, a demonstration employing a racing shadow is effective.

APPARATUS NEEDED: The apparatus needed to produce the effect of this motion consists of a wooden disk of some convenient size that can be rotated smoothly about its central axis. Located on the periphery of the disk is a vertical wooden dowel. When the disk is rotated in a beam of bright light,[2] the shadow cast by the rotating wooden dowel will move back and forth across a screen or light-colored wall.

The observer watching this shadow actually sees an example of simple harmonic motion. Since the relative size of the shadow and length of its run depend upon the relative distances between the screen, the disk, and light source, the projected size of the demonstration can be easily adjusted.

Principal Objective

The principal objective of this demonstration is to show that the rate of

---

[1] Simple motion is defined as motion along a straight line or circular arc.

[2] A slide projector makes a good light source.

change of the shadow's velocity, its acceleration that is, is proportional to the displacement of the shadow from the center of the screen, but in the opposite direction.

In order to visualize what this means physically, let us examine how the shadow's *displacement, velocity,* and *acceleration* each change during one trip of the shadow across the screen and back. If we let (X) and $(-X)$ be the length of the shadow's respective displacement to the right and left of center, and let $(-X_1)$, (X_0), and (X_1) be the respective shadow positions at the extreme right, extreme left, and center of the path, we can graphically represent the displacement, velocity, and acceleration of the shadow as it moves from one position to another during its round trip (Fig. 1).

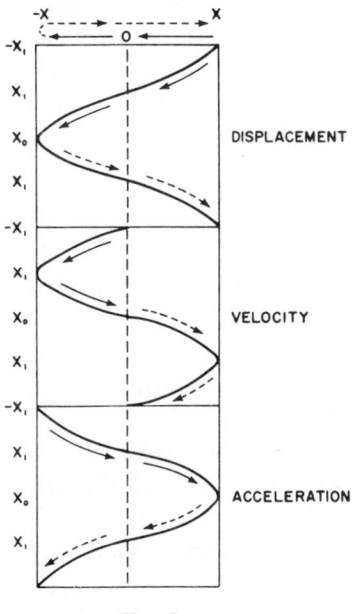

Fig. 1

Although the shadow can begin its trip across the screen from any position, let us have the shadow begin from the extreme right. At this location, the displacement is positive. As the shadow begins to move toward the center of the screen, the displacement decreases in value, but remains positive. The velocity of the shadow increases in a negative direction, reaching a maximum at the center of the screen; the acceleration, which has a maximum value at the moment the shadow begins to move, decreases as the shadow approaches the center. Because the rate of change of velocity is in a negative direction, the acceleration is negative.

NOTE: If we look at the shadow when it is exactly at the center position, we see that its displacement is now zero, but its velocity has reached a maximum. Since no change in velocity is occurring at this point, the acceleration must be zero the instant the shadow is at the center position.

As the shadow moves past the center position and toward the left side of the screen (negative displacement), the velocity decreases. But since the velocity decreases in a negative direction, the rate of change of velocity is positive and so the acceleration is positive. Negative deceleration is equivalent to positive acceleration. When the shadow arrives at the extreme left of the screen, it stops, then reverses direction and heads back toward the center of the screen. Now the change in displacement is in a positive direction; however, it (the displacement) still has a negative value.

Because the velocity of the shadow at the extreme left of the screen changes from a negative value to a positive value, the acceleration is positive and at a maximum initially. The shadow's velocity increases during its return to the center of the screen but the acceleration decreases, becoming zero at the center of the screen. Although the acceleration decreases as it approaches the center, it retains a positive value until the shadow is in the exact center of the screen.

Finally, as the shadow returns to its original position at the far right of the screen, the displacement increases positively in both direction and value. The velocity decreases in a positive direction, from a maximum at the center to zero at the edge so the rate of change of velocity is negative, or in other words, the acceleration is negative.

Summary

At each of the shadow locations we examined, we saw that the acceleration was always directly proportional to the displacement. When the displacement was a maximum, the acceleration was a maximum; when the displacement was a minimum, the acceleration was a minimum. We also saw that the direction of the displacement was always in a direction opposite to the acceleration. Since these are the criteria of simple harmonic motion, we can feel confident that the actual motion we saw was in fact simple harmonic motion.

NOTE: This demonstration can also be used to introduce or review the need for the use of vector notation. Recall that at each shadow position, we had to consider both the value (magnitude) and direction of the displacement, velocity, and acceleration.

9

CONSERVATION OF
MOMENTUM

DEMONSTRATING THE DIFFERENCE
IN FORCE ON A REFLECTING AND
AN ABSORBING SURFACE

by Arthur Brachmann

Forest Grove High School, Forest Grove, Oregon

Following is a good way to demonstrate that to stop an object and get it going at the same speed in the opposite direction takes twice as much energy as it does to merely stop the object. The idea for this demonstration comes from the PSSC film, "The Pressure of Light," in which Dr. Zacharias shoots bullets at a metal block.

Materials

PSSC cart
¾" steel ball
2" x 4" block of wood about 8" long
crutch tip
ball ramp (Macalester's Rotational Dynamics Apparatus)

Procedure

1. Place the block of wood on the cart and nail cleats on the cart to prevent the block from sliding backward and forward.
2. Attach the crutch tip to one end of the block so that the steel ball rolling down the ramp will catch in it when the block is on the cart. (Since the block is not nailed on the cart, it can be easily turned around.)
3. Fasten the ball ramp to a smooth, level surface 1 inch by 12 inches by about 4 feet long.
4. Place the cart in a position—with block and crutch tip in location—so that as the ball leaves the ramp it rolls into the crutch tip.
5. Release the ball down the ramp so that it rolls into the crutch tip.
6. Measure the distance the cart moves (Fig. 1).
7. Then, place the cart in the same position with the block reversed so the ball will bounce off of it.
8. Release the ball from the same place on the ramp so that each time it leaves the ramp it possesses the same kinetic energy.

NOTE: The cart will roll twice as far when the ball bounces off of it as when the ball sticks to it (Fig. 2).

Fig. 1

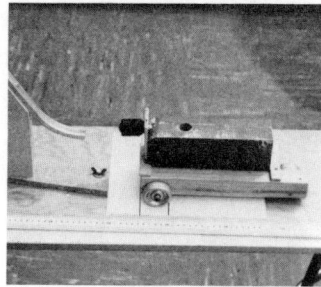

Fig. 2

STUDENT-CONSTRUCTED DEMONSTRATION OF NEWTON'S THIRD LAW

by Paul A. Wilkinson

Manuel High School, Denver, Colorado

This device helps general science students understand Newton's Third Law: For every action there is an equal and opposite reaction.

Materials

Ring stand
Bunsen burner
Tin can, with lid (such as found on baking powder cans)
Thread
Screw eye bolt

Demonstration

Solder screw eye bolt to lid of can. Dent the side of the can slightly, so that a portion of the wall is at an angle to the generally cylindrical surface. Punch a hole into this indentation so that steam will come out of the hole in a path as nearly parallel to the side of the can as possible (Fig. 1).

Fill can one-quarter full of water and attach lid *tightly*. Heat bottom of can with burner. If you use a Bunsen burner, the can will start rotating within one or two minutes, depending on the size of the can, initial temperature of the water, and the amount of water in the can. After rotating, when heat is removed, the can will start rotating in the opposite direction, caused by the unwinding of the string (this latter, is not part of the Third Law).

The demonstration attracts the interest of the class. The fact that it is student performed does not answer the question, "Why does it spin?" But it does encourage classroom questions and discussions, guided by the teacher.

NOTE: The experiment is a modification of a steam toy made by Hero, a scientist of ancient Alexandria.

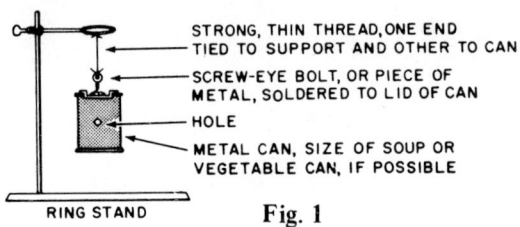

STRONG, THIN THREAD, ONE END TIED TO SUPPORT AND OTHER TO CAN

SCREW-EYE BOLT, OR PIECE OF METAL, SOLDERED TO LID OF CAN

HOLE

METAL CAN, SIZE OF SOUP OR VEGETABLE CAN, IF POSSIBLE

RING STAND

Fig. 1

10

WORK AND ENERGY

INTRODUCING THERMODYNAMICS

by John M. Hicks and Robert N. Grant

Alamo Junior High School; Goddard Junior High School,
Midland, Texas

Here is a short, three-part laboratory block designed to introduce some of the basic concepts of thermodynamics at the ninth grade level. Part I is a guided discovery of the heat of fusion, thermal capacity, and heat of vaporization of water. Part II is an activity planned to give students some insight into the meaning of entropy. The last part is a laboratory verification of the first and second laws of thermodynamics.

Materials:

1 large, 2 small styrofoam coffee cups
Electric immersion heater (Japanese-made)
Thermometer
Watch with a sweep second hand

I. Phase Changes and Thermal Capacity of Water

Students should understand that, like all compounds, water can exist as solid, liquid, or gas; that, at constant pressure, we normally identify these three phases with temperature ranges. Thus, if an electric immersion heater is placed in a styrofoam cup of crushed ice, the heater will work to melt the ice. It will change the temperature of the liquid water, and then boil the water, producing steam.

Explain that since work delivered by a constant power is proportional to time, we may, by means of a temperature vs time graph, determine how much work is required: (1) to melt a unit mass of ice (heat of fusion), (2) to change the temperature of a unit mass of liquid water 1 Kelvin degree (thermal capacity), and (3) to change a unit mass of liquid water to steam (heat of vaporization).

Procedure:

1. Place a known mass of crushed ice, a thermometer, and an immersion heater in a styrofoam cup.
2. Connect the heater and start the timer.
3. Record the temperature every 15 seconds.

192

4. Indicate at what time: (a) the ice melted, (b) the water boiled.
5. After the water has boiled for about five minutes, disconnect the heater.
6. Calculate the mass of steam produced.
7. Graph temperature vs time for the heating of water in a space below (Fig. 1).
8. Calculate the heat of fusion, thermal capacity, and heat of vaporization for the water.

Questions:

(a) How much work was done on the water by the heater?
(b) What part of the work went to: (a) melting the ice? (b) changing the temperature of the water? (c) vaporizing the water?
(c) Calculate the heat of fusion of ice.
(d) Calculate the thermal capacity of water.
(e) Calculate the heat of vaporization of water.
(f) How do your results compare with the average obtained by the class?

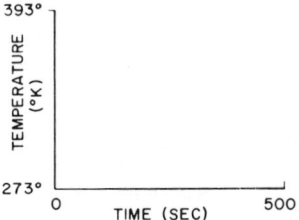

Fig. 1

II. The Concept of Entropy

Students should be aware that a convenient measure of the status of water is proportional to the probability of the arrangement of its molecules, that another measure is proportional to the average molecular velocity. The first of these is called "entropy" and the second "temperature." They should understand that changes in both are related to, and vary directly with, the work done (heat added) to the water ($\triangle Q \mathbb{C} T \triangle S$, where S represents entropy, T the absolute temperature, and Q the quantity of heat).

Entropy may be explained as the measure of the degree of disorderliness, or mixed-up-ness, of a system. For example, the entropy of 100 coins, all heads (the least probable distribution), is zero. In the case of 100 coins, 50 heads and 50 tails (the most probable distribution) is a maximum. Work is equal to force x distance, pressure x volume, voltage x charge, or temperature x entropy. The area under a graph of the factor which does not depend upon the amount of material vs the quantitative factor—which does depend on the amount of the material—is equal to work.

Point out that in Part I, since the work done is proportional to time, the temperature vs time graph may be directly converted to a temperature vs work graph. From the relation$\Delta Q \subset S$, which converts to $\Delta S \underline{\Delta Q}$ with the proper
$$\text{T}$$
units, it is obvious that an entropy vs temperature graph would be symmetric with the temperature vs heat graph.

Procedure:

1.   Sketch a temperature vs entropy graph for the heating of water (Fig. 2)

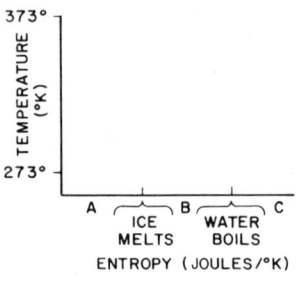

Fig. 2

2.   On the graph, near the middle of section B, shade in an area with parallel, vertical sides that represents the work required to change the temperature of all the water 10 Kelvin degrees.
3.   Shade in rectangles under sections A and C of the graph with area equal to that shaded in Section B.

Questions:

(a)   How many joules are represented by the shaded area?
(b)   How do the distances between the vertical parallel lines (the change in entropy) compare for each section?
(c)   How do the average values of the temperature at which the work is done to produce the changes in entropy, compare?
(d)   For a given amount of work, how does the change in entropy or disorder vary with the temperature at which the work is done?

III. The First Law of Physics and the First Law of Nature

Explain that changes in entropy are smaller the higher the average temperature at which the work, which produced the entropy change, takes place $(\Delta S = \underline{\Delta Q})$. Energy is transferred by work, and heat energy in a given mass is
$$\text{T}$$
changed as the average molecular velocities (temperature) change.

Introduce the two laws which govern all physically real processes:

— The energy of the universe is constant.

— The entropy of the universe tends toward a maximum.

The first of these laws is clearly revealed in the physics laboratory where we constantly apply it as input equals output, heat lost equals heat gain, etc. The second is encountered in all of nature; for instance, broken windows do not unbreak; scrambled eggs do not unscramble; clocks run down; etc.

Point out to students that if a cup of hot water is mixed with a cup of cold water, we can predict from analogous experiences that the temperature of the mixture will lie between the temperature of the two original cups. However, if we apply the two laws, we can predict exactly what the final temperature will be providing we know the initial temperature and masses.

Procedure:

1. Pour a volume of hot water into a small styrofoam cup and an equal volume of cold water into a smaller cup.
2. Record the temperatures and masses of water in each cup.
3. Mix the two cups of water together in a large styrofoam cup and record the temperature of the mixture.
4. Calculate the changes in energy for both the hot and cold water.
5. Calculate the changes in entropy for both the hot and cold water.

Questions:

(a) How much energy was gained or lost when the water was mixed?

(b) How much net entropy was gained or lost when the water was mixed?

(c) Based on the evidence provided by this experiment, what do you predict is the ultimate fate of the universe?

(d) If entropy had been conserved in mixing the water, what would have happened to the energy?

UNDERSTANDING THE CONCEPTS OF CONSERVATION OF ENERGY AND WORK

by Phillip Rosete

University of Florida, Gainesville, Florida

In teaching physics, we strive to bring about as complete an understanding of a concept as possible. To illustrate, here is a rather simple and inexpensive way to set up an experiment that is geared to give the student an understanding of the basic concepts of conservation of energy and work.

This experiment is used most effectively if PSSC Physics is taught. By the time the experiment is performed, the students have become familiar with the rudiments of: (a) interpretation of graphs, (b) potential and kinetic energy, and (c) work. Among other things, they have made use of the PSSC hand strobe; they have learned to extrapolate and interpolate; interpreted information in basic graphs, such as the graph of velocity vs time to find distance and acceleration; and have been introduced to the techniques of finding the area under curves. Thus, the experiment is effective because it makes use of apparatus and techniques with which they are familiar.

The Apparatus

The main part of the experiment consists of a plunger assembly made from an old PSSC wooden collision cart. It is a piece of wood 2" x 4" x 9", with a 5/8" hole bored through it. A spring is inserted in the hole and secured at one end with a wood screw. A piece of 1/2" electrical metal tubing (which was a part of the original cart) is attached to the loose end of the spring, and a wooden stopper is inserted at the other end of the metal tubing. (We tried a rubber stopper, but found the string in the device cut through it.) (See Fig. 1.)

Other materials needed are:

PSSC timer kit
PSSC strobe
Two .5 kg hooked standard masses
Two 1 kg hooked standard masses
Wooden incline board with pulley

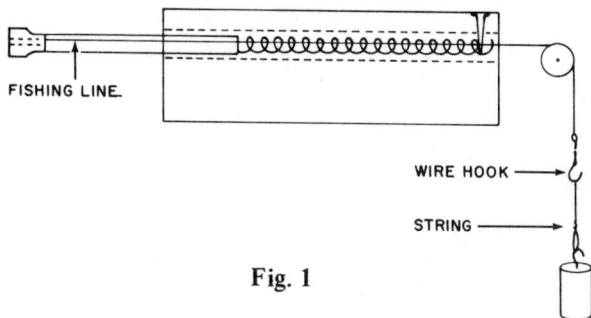

FISHING LINE.

WIRE HOOK ——

STRING ——

Fig. 1

PSSC collision cart (any cart)
C-clamps, matches, fishing line, meter stick, wire hooks

Secure a small piece of fishing line to the stopper through two small holes drilled in the stopper. Bring the line through the metal tubing, through the inside part of the spring, and out the rear of the assembly. (The wooden screw is placed off-center to avoid contact with the line.) A small hook is tied to the end of the line and the line is placed over a pulley to the rear of the assembly. A hooked standard mass is suspended by means of a small string to the wire hook. This will provide the constant force which will compress the spring. To release the spring it is only necessary to apply a burning match to the string. A cart properly placed against the wooden stopper will be accelerated during the interval of time that the spring is returning to normal length. A paper timer attached to the cart by a piece of masking tape and running through a timer will record this motion.

The following procedure is suggested to the students:

1. Set up the experiment as shown in Fig. 2.
2. Suspend a .5 kg mass and record the amount of compression$\triangle$X.
3. Increase mass to 1.0 kg, 1.5 kg, and 2.0 kg. Record the compression in each case.
4. Accelerate the cart when the 2.0 kg mass is suspended from the string.

In this experiment, we tell the students what to do, but not how to do it. It is designed with the purpose of suggesting things that the student should do. While all may not do so, many may. To illustrate:

(a) In order to plot a graph of Potential Energy vs Compression, the student may well be forced to plot a graph of Force vs Compression and render a proper interpretation. Reason: Since the force applied to compress the spring can be found by the suspended mass times the acceleration due to gravity and since the compression of the

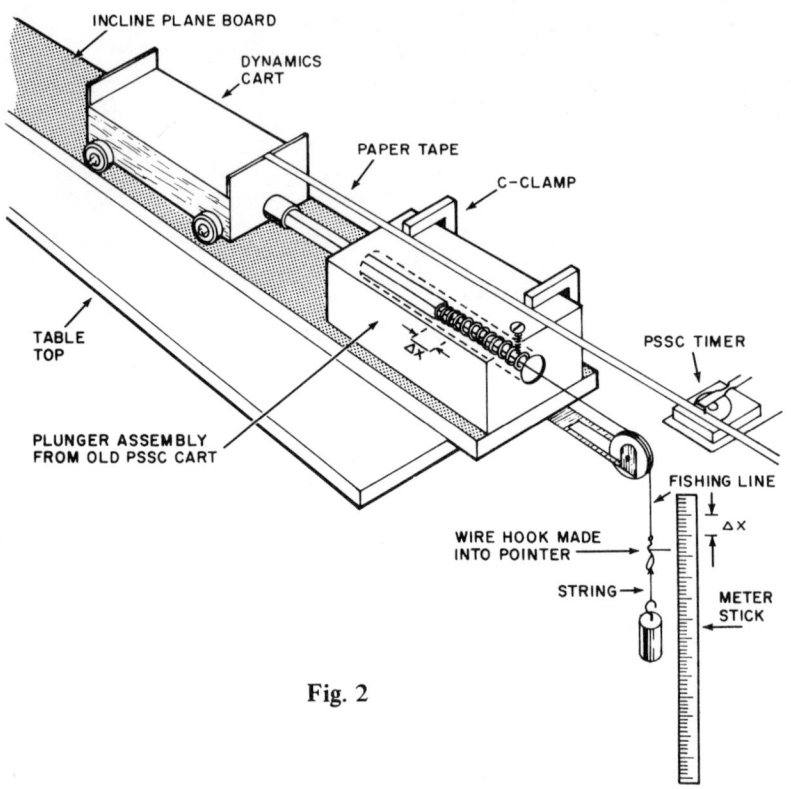

Fig. 2

spring can be measured, then a graph of Force vs Compression would be useful in finding the potential energy stored in the system.

NOTE: There are other ways to find this potential energy, of course, but a graph is more meaningful at this stage.

(b) In order to plot the graph of Kinetic Energy vs Compression (the kinetic energy of the system), the student must think back and realize that he must compute the final velocity of the moving part of the system. This, in turn, will almost certainly force him into making use of some type of calibrated timing device. The actual use of the hand strobe in this situation helps to develop a meaningful understanding of the strobe.

We do not stress a final correct answer to some number of significant digits. The answer must be correct to a certain number of digits, but we might accept accuracy to two digits instead of three if the order of magnitude is correct. What we are looking for is the student's approach in using the available data and instruments at his disposal to the fullest advantage.

Typical Data and Graphs Plotted

Following (Fig. 3) is a table obtained by using a cart of mass = 1.740 kg and a timer whose frequency = 48.3 vibrations/second.
Using the suggested procedure above, following is the data collected:

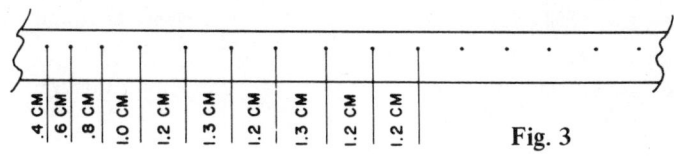

Fig. 3

	M Mass	$\triangle$X Spring Compression	F Force on Spring
1.	0.0 kg	0.0 m.	0.0 nt
2.	0.5 kg	1.2×10^{-2} m	4.9 nt
3.	1.0 kg	2.5×10^{-2} m	9.8 nt
4.	1.5 kg	3.9×10^{-2} m	14.7 nt
5.	2.0 kg	5.1×10^{-2} m	19.6 nt

$F = MA$
$F = .5 \text{ kg} \times 9.8 \text{ m/sec}^2$
$f = 4.9 \text{ nt}$

To plot graph of Potential Energy: Since Potential Energy is a measure of the work put into the system, it may be found by the area under the curve of Force vs Compression. (See graph, Fig. 4.)

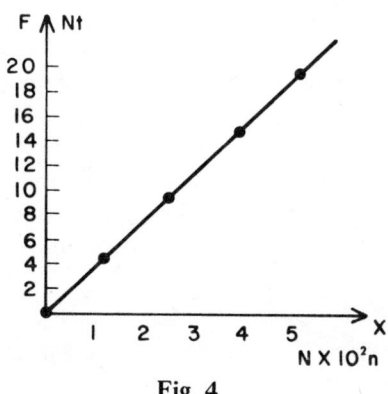

Fig. 4

$$U_1 = \frac{1}{2} FX$$
$$U_1 = \frac{1}{2} (0)(0) = 0.0 \text{ joules}$$
$$U_2 = \frac{1}{2} (4.9 \text{ nt})(1.2 \times 10^{-2} \text{ m}) = 2.9 \times 10^{-2} \text{ joules}$$
$$U_3 = \frac{1}{2} (9.8 \text{ nt})(2.5 \times 10^{-2} \text{ m}) = 12.3 \times 10^{-2} \text{ joules}$$
$$U_4 = \frac{1}{2} (14.7 \text{ nt})(3.9 \times 10^{-2} \text{ m}) = 28.7 \times 10^{-2} \text{ joules}$$
$$U_5 = \frac{1}{2} (19.6 \text{ nt})(5.1 \times 10^{-2} \text{ m}) = 49.9 \times 10^{-2} \text{ joules}$$

To plot a graph of Kinetic Energy: It will be necessary to find the final velocity of the cart at the end of each interval of time during the decompression of the spring as given by the recorded ticks on the tape.

Using the PSSC strobe, the frequency of the timer is found to be 48.3 vib/sec. The period is then $1/48.3$ vib/sec $= 2 \times 10^{-2}$ sec/vib.

The final velocity at the end of the first interval of time: $V = s/t = .4 \times 10^{-2}$ m$/2 \times 10^{-2}$ sec $= .2$ m/sec.

And the K.E. will be: K.E. $= \frac{1}{2} MV^2 = \frac{1}{2} (1.74 \text{ kg}) (.04 \text{ m}^2/\text{sec}^2) = .035$ joules.

Following is the data for plotting the kinetic energy:

Time Int.	Dist. Covered	Compression of Spring	Final Velocity	Kinetic Energy
0	0	5.1×10^{-2} m	0.0 m/sec	0.0 joules
First	$.4 \times 10^{-2}$ m	4.7×10^{-2} m	0.2 m/sec	0.035 joules
Second	$.6 \times 10^{-2}$ m	4.1×10^{-2} m	0.3 m/sec	0.08 joules
Third	$.8 \times 10^{-2}$ m	3.3×10^{-2} m	0.4 m/sec	0.14 joules
Fourth	1.0×10^{-2} m	2.3×10^{-2} m	0.5 m/sec	0.22 joules
Fifth	1.2×10^{-2} m	1.1×10^{-2} m	0.6 m/sec	0.31 joules
Sixth	1.3×10^{-2} m	0.0 m	0.65 m/sec	0.37 joules

A plot of the Total Energy vs Compression yields the graph shown in Fig. 5.

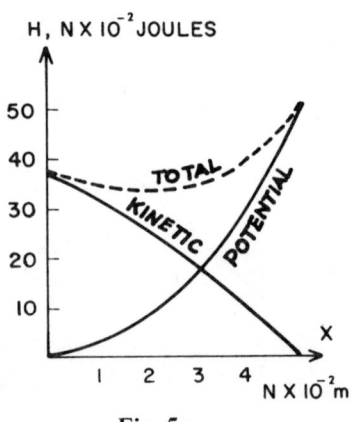

Fig. 5

For a conclusion to the exercise, the students may consider the following questions:

1. Is the graph of Potential Energy what you expected it to be?
2. Is the graph of Kinetic Energy what you expected it to be?
3. Did we actually get back all of the energy that was put into the system? Be specific.
4. What is the meaning of the point of intersection of the Potential and Kinetic Energy graphs? Be specific.
5. Explain the dip on the curve of the Total Energy.
6. Was the Total Energy really conserved?
7. What was the total Potential Energy stored in the spring at maximum compression?
8. What was the Kinetic Energy of the cart at the time of separation from the plunger?
9. If the answers to No. 7 and No. 8 are equal, what is your explanation?
10. If the answers to No. 7 and No. 8 are not equal, what is your explanation?

NOTE: The tables, charts, graphs, and results shown here are to serve as a guide of what may be expected from students. In the actual experiment, of course, they are not given as a requirement in the instructions; this would eliminate much of the challenge to the student.

AN ANALYSIS OF MECHANICAL ENERGY

by Chris Buethe

New Mexico State University, Las Cruces, New Mexico

Familiar objects used in unfamiliar ways can present interesting challenges to students. Such is the case with the following physics experiment, one that is based upon an "energy approach" to problem solving.

NEWTON'S LAW: After being introduced to Newton's second law and the laws of conservation of momentum and energy, unsophisticated physics students are confronted with the apparatus shown in Fig. 1.

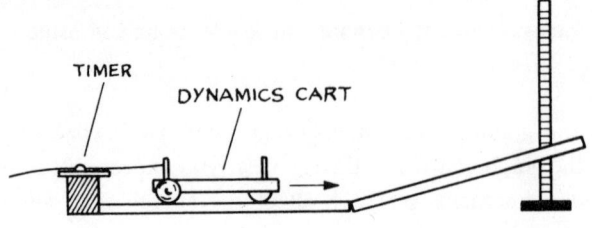

Fig. 1

Their instructions call only for them to "cock" the cart mechanism, then account for energies as the triggered cart is pushed and finally rolls up the ramp to its maximum height.

Apparatus and Materials

Per Station

    Dynamics cart
    Recording timer
    Spring balance
    Meter stick
    Ring stand
    Clamps
    Wood block
    Two smooth 1" x 6" inclines (approximately 3' each)

Per Class

    Stopwatch (to calibrate timer)
    Bathroom scale

    NOTE: Students really see that the bathroom scale is useful in determining the energy that can be stored in the spring. Good questions about energy represented by the area under a curve are likely to follow. It is very tempting to overinstruct at this point.

Some students are disturbed by the $2\pm$ joules of potential energy stored in the spring that reduce to the cart's $1\pm$ joule of kinetic energy—and finally, to the cart's $1\pm\dfrac{}{3}$ joule of gravitational potential energy.

The student with sufficient insight will account for energy "losses," usually by measuring force x distance products by means of the spring balance and meter stick.

This rather open-ended experiment incorporates the primary principles of Newtonian mechanics, lending itself to a spectrum of student response levels. Grading can be based, in part, upon how well students indicate an understanding of measurement, conversion of units, graphic representation of data, and the concept of center of mass. The experiment serves well to introduce a discussion of nature's tendency to reduce order.

FORCE PARALLEL TO DISPLACEMENT—
WORK AND ENERGY: AN EXPERIMENT

by Max Hamilton

Hutchinson Senior High School, Hutchinson, Kansas

I find this experiment most interesting and educational as it involves a great deal of physics in a very simple way.[1] Before the teacher assigns the experiment to his students, however, it is important that he try each step of the procedure himself.

Materials: Necessary equipment includes a hack-saw blade (or a PSSC spring blade), 2 C-clamps, a piece of string, a small strip of masking tape, and several nickels. The nickels are used because they weigh almost exactly .05 newtons each.

Fig. 1 shows a rough sketch of the assembled apparatus. A lab desk drawer supplies the basic framework for the setup. A small block of wood is clamped to one side of the drawer near the top, then the blade is clamped to the block in a position similar to that of a swimming pool springboard. To one side of the blade and to the bottom of the drawer, a piece of graph paper is fastened, so that a distant light will cause the blade to cast a shadow on the paper.

> **PROCEDURE: Support the blade with a C-clamp and adjust the piece of graph paper adjacent to it so that the distant light source will cause a shadow of the blade on the paper. Tie a string to the loose end of the blade, and place a strip of masking tape (to which the nickels will later be attached) on the string.**

Qualitative Description

When the students have set up their apparatus, they may be given laboratory worksheets on which to record their experimental data. They may be asked first to supply the qualitative information called for in the following:

---

[1] The idea for this experiment was given to me by Dr. Gerald Witten, Emporia State College, Emporia, Kansas.

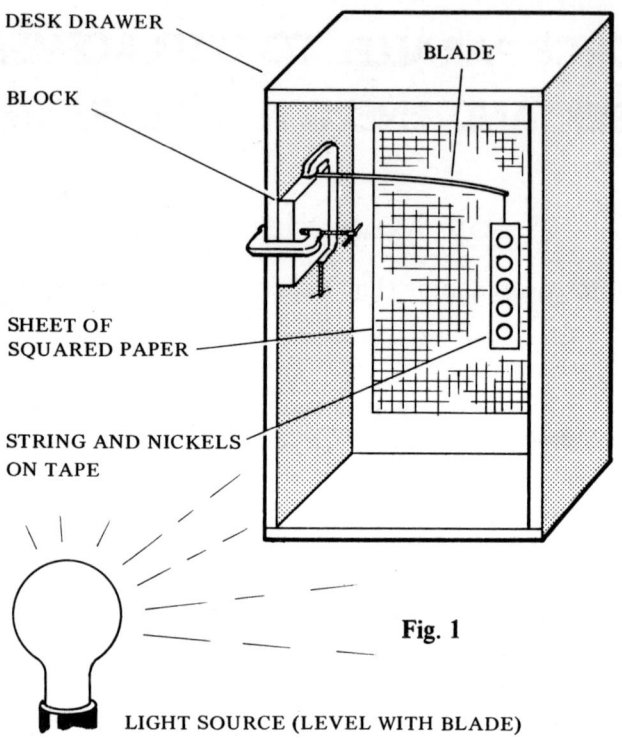

DESK DRAWER

BLADE

BLOCK

SHEET OF
SQUARED PAPER

STRING AND NICKELS
ON TAPE

Fig. 1

LIGHT SOURCE (LEVEL WITH BLADE)

(1) As a force moves through a distance parallel to the distance, work is done. A 1-newton force moving through a distance of 1 meter does 1 joule of work, provided that the 1-newton force is parallel to the 1-meter force. Pull on the string attached to the blade. Is a force required to distort the blade?_____.

(2) Compare the force needed to distort the blade from 0 to 1 centimeter to the force required to distort the blade from 1 to 2 centimeters. The latter force is _____ .

(3) Considering that the displacement is the same in both cases (1 centimeter), then the greater amount of work is done in the first or the second case?_____.

(4) When the blade is distorted 2 centimeters, is energy stored in the blade? _____. Is the blade able to work for you?_____.

(5) Place a small paper wad on the end of the blade and release the blade. What energy transformations do you see? (student statement)_____.

(6) Place five nickels on the masking tape which is attached to the string, which is attached to the blade. Note the rest position—_____ cm. Pull the nickels down about 1 or 2 centimeters (low or limit) and release. Notice that the blade goes up to an upper limit.

Upper Limit is _____ cm.
Rest Position is _____ cm.
Lower Limit is _____ cm.

The blade has maximum energy stored in it at (UL, LL, or RP)_____ .

The blade has least energy stored in it at_____ .

The nickels have most gravitational potential energy when the blade is at _____.

The kinetic energy of the nickels is greatest when the blade is at_____ .

Quantitative Description

Students may then be asked to complete a quantitative description of the experiment in a table supplied on their laboratory sheets (Table 1). Finally, they can plot a graph of force in newtons on the y axis vs distortion in meters on the x axis, and furnish the information called for from an examination of their graphs.

Table 1

Number of nickels	0	1	2	3	4	5	6
Mass (grams)	0	5	10	15			
Force (newtons)	0	5×10^{-2}	10×10^{-2}	15×10^{-2}			
Distortion (meters)	0						

(a) *Look at your graph.* The slope of the line is _____ nt/m.

(b) The work done to distort the blade 5 cm $(5 \times 10^{-2}$ m) is _____ joules. *Be careful.* The final force is_____ nt and the initial force is_____ . Therefore, the average force is_____ nt and the work done would be_____ joules to distort the blade 5 cm. *Get these answers from your graph.*

(c) *Look at your graph.* What is on the y axis?_____ . The x axis measures what quantity?_____ . What is the product of force and distance?_____ . Could the area under the curve represent work? _____ . Why? (student statement)_____ .

(d) *From your graph,* the change in potential energy stored in the blade as it is distorted from UL to LL is_____ joules. Using $Epg = mgh = wh$, the change in potential energy of the nickels due to the change in position in the gravitational field is_____ joules from UL to LL. How do these answers compare?

(e) *From your graph,* the change in energy stored in the blade from RP to UL is_____ joules. Using mgh, the change in gravitational potential energy of the nickels from RP to UL is_____ joules.

(f) Can you explain why the two answers to (e) are not equal? Try.

11

STATIC ELECTRICITY

DISCOVERING THE CHARGE-DISTANCE RELATIONSHIP FOR A CONSTANT FORCE

by George F. Smith

South Hadley High School, South Hadley, Massachusetts

This is an experiment to discover the charge-distance relationship between two like charges when the repulsive force is constant.

APPARATUS: The charges are carried on lightweight, plastic spheres[1] coated with graphite mounted on thin slivers of insulating plastic. One sphere is mounted on the short end of a soda straw balance (Fig. 1). Two other identical spheres are mounted on corks—one is attached to a ring stand and the other is a spare (Fig. 2). Using a ring stand, light source, meter stick, and the soda straw balance, set up as in Fig. 3 to measure shadows of the plastic spheres on the meter stick.

Procedure

Using a charged rod or a PSSC-type charged plastic strip, charge by induction both the supported sphere and the one on the balance with the same kind of charge. They should repel each other. Record the positions of the two sphere shadows on the vertical meter stick. Touch the spare sphere to the identical supported one to remove some of the charge, and give this charge to the ground by touching the spare sphere with a finger.

NOTE: The charge left on the supported sphere will be just about half of what it was originally, since by symmetry the charges should share 50-50. If the spheres were not the same size then it would be rather difficult to determine the charge distribution.

Move the supported sphere vertically so that the bottom shadow is in the same position as it was before and record the new position of the upper shadow. Compute by subtraction the ORIGINAL SEPARATION DISTANCE (R) and the FINAL SEPARATION DISTANCE (R').

---

[1] Macalester Scientific Corporation #MSC 122, ½-inch pith balls coated with #MSC 1085 graphite in alcohol.

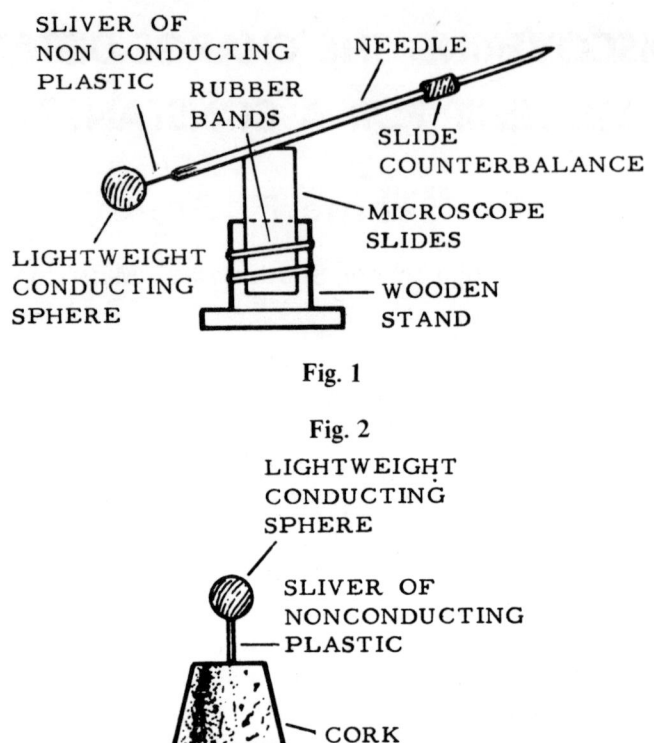

Fig. 1

Fig. 2

Plot the Data

Repeat this procedure about a dozen times using different starting positions for the lower shadow. Plot the data on a graph of FINAL SEPARATION DISTANCE (R') vs ORIGINAL SEPARATION DISTANCE (R). Each trial gives a new (R, R') point on the graph. Ideally, the magnitude of the ORIGINAL SEPARATION DISTANCES (R) should be varied to give a good spread in the data points. Plotting the data points as soon as they are determined will give the experimenter a better idea of what magnitude ORIGINAL SEPARATION DISTANCE (R) to use on succeeding trials in order to get a good graph.

The ratio (q'/q) of the final charge (q') on the supported sphere to the ORIGINAL CHARGE (q) on the same sphere should be ½. The ratio (R'/R) between the FINAL SEPARATION DISTANCE (R') and the ORIGINAL SEPARATION DISTANCE (R) can be determined by finding the slope of the graph just plotted. The problem is to see how the ratio q'/q compares to the ratio R'/R. With this information, a student should be able to determine how the charge (q) varies mathematically with the separation distance (R) if all other variables are kept relatively constant.

ANSWERS: From reasonably good students, answers like $q = KR^2$,

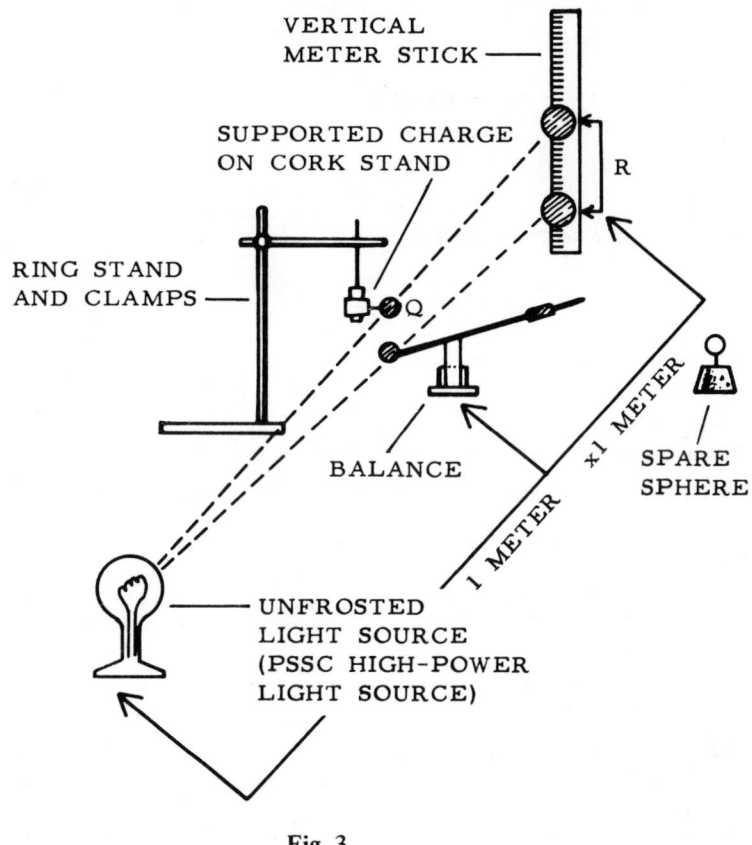

VERTICAL
METER STICK ——

SUPPORTED CHARGE
ON CORK STAND

R

RING STAND
AND CLAMPS ——

Q

BALANCE

SPARE
SPHERE

1 METER

x1 METER

—— UNFROSTED
LIGHT SOURCE
(PSSC HIGH-POWER
LIGHT SOURCE)

Fig. 3

$q'/q = R'^2/R^2$, and $\sqrt{q'/q} = R'/R$ should be expected. The rest of the class will catch on soon afterwards.

Coulomb's Law

In this experiment, the governing relationship is Coulomb's Law, $F = kQq/R^2$. If F, k, and Q remain constant during any trial, then $q = KR^2$ or $\sqrt{q'/q} = R'/R$. Since in each trial the charge, q, is cut in half, $\sqrt{q'/q} = R'/R = \sqrt{\frac{1}{2}} = 1/1.4$. The exact placement of the apparatus to produce shadows is not important and in fact may be changed between trials. It is recommended that different forces (starting positions for bottom shadow) be used for each trial to obtain separation of data points on the graph.

SAMPLE GRAPH: Leakage should not be any problem since any one trial takes about 30 seconds. The data for the sample graph

Chart I

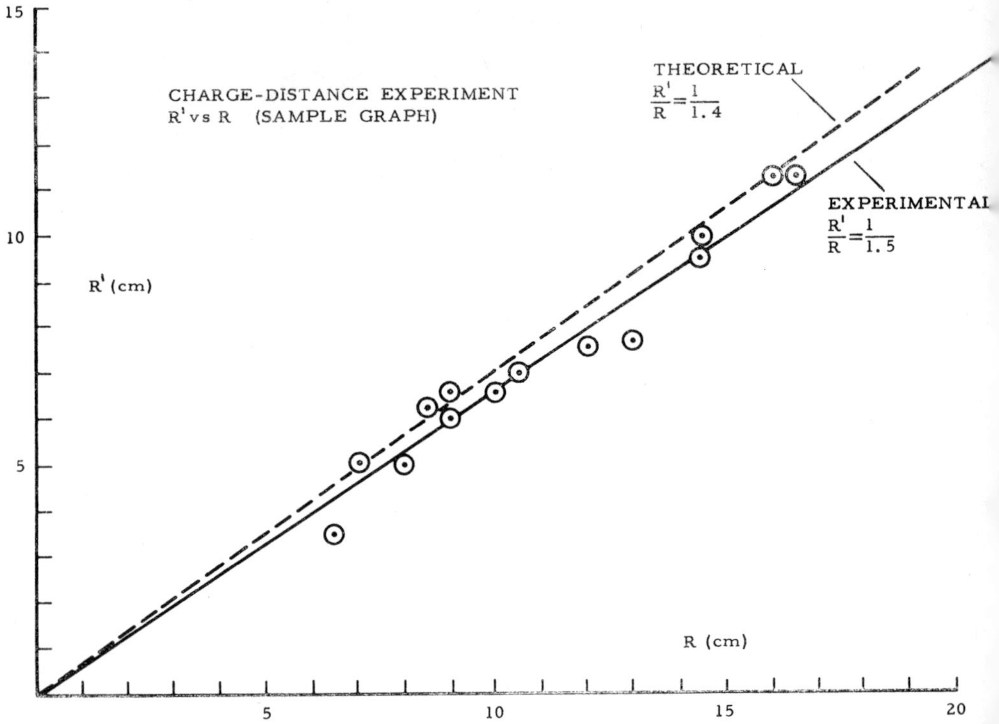

CHARGE-DISTANCE EXPERIMENT
R'vs R   (SAMPLE GRAPH)

THEORETICAL
$\frac{R'}{R} = \frac{1}{1.4}$

EXPERIMENTAL
$\frac{R'}{R} = \frac{1}{1.5}$

R' (cm)

R (cm)

(Chart I) was obtained under conditions of extremely high humidity. If necessary, shielding of the balance from air currents can be achieved in many ways, including putting the balance into a dry fish tank.

This experiment makes a suitable demonstration-experiment when a class is learning about Coulomb's Law or a good optional experiment for talented students. Average students have a little difficulty with the required delicacy of the apparatus but eventually are successful.

NOTE: Inspiration for this experiment came from the PSSC film "Coulomb's Law" starring Eric Rogers of Princeton University.

LABORATORY PLOTTING OF
ELECTRIC FIELDS

by Lawrence B. Ryan

Beaver River Central School, Beaver Falls, New York

The plotting of electric lines of force with a pair of headphones is a typical laboratory exercise in modern high school physics courses. Students use the headphones to locate equipotentials in the electric field between various-shaped electrodes in a mildly conducting saline solution. The field is established by connecting the electrodes to the secondary terminals of a low-voltage transformer, as shown in Fig. 1.[1]

NOTE: Use of headphones limits this "null deduction" exercise to one student at a time. The adaptations described here will allow the entire class to participate in the exercise together.

Adapting the Setup

Since the resistance of a few centimeters of a weak saline solution is a few thousand ohms, feed the signal from the detecting probes directly to the high-impedance input of a good quality audio amplifier as shown in Fig. 2.

CAUTION: The amplifier should be transformer operated for safety.

The amplified signal, fed through the loudspeaker, is audible to a large group, and the detection of null points can be made very sensitive by increasing the amplifier gain.

A possible drawback to this is that many amplifiers available in small school labs have relatively high "hum" levels; even with no signal input, they may produce an audible 60 Hz signal at the output. This "hum" is the same frequency as the alternating potential that establishes the field under measurement and can interfere with the detection of equipotentials.

To avoid the "hum," use an alternate field source. Instead of the output of a transformer, employ an audio oscillator (Fig. 2) to provide an alternating emf

---

[1]Fig. 1 is redrawn from page 231 in Turner and Carpenter's *Discovery Problems in Physics* (New York: College Entrance Book Co., 1966).

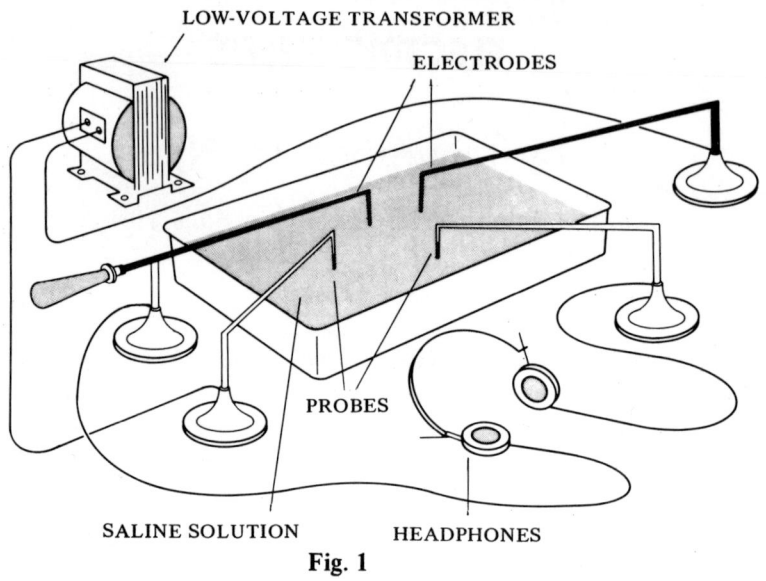

LOW-VOLTAGE TRANSFORMER

ELECTRODES

PROBES

SALINE SOLUTION   HEADPHONES

Fig. 1

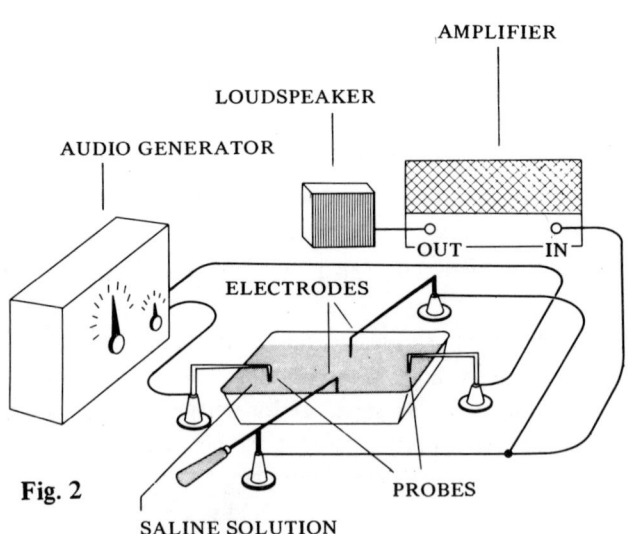

AMPLIFIER

LOUDSPEAKER

AUDIO GENERATOR

OUT — IN

ELECTRODES

Fig. 2

PROBES

SALINE SOLUTION

of any convenient frequency. When set to from 500 to 1,000 Hz, the generator establishes a field in which null points are easily detected.

DEMONSTRATING STATIC ELECTRICITY

by William J. Muha

Formerly Notre Dame High School, Niles, Illinois

Here are three simple demonstrations of static electricity which I have found useful with my physics students. The first two are effective as teacher demonstrations to rouse student interest and curiosity. The third will give students an opportunity to conduct their own demonstrations using homemade "electroscopes."

Dramatizing Static Electricity

For a dramatic display of static electricity, obtain several ordinary balloons from a local variety store and inflate them with air.

(a) Suspend one balloon with two strings, so that the strings form a "V" with the balloon at the vertex.

(b) Tape the second balloon to the end of a meter stick in a fixed position.

(c) Rub both balloons with fur, then try to bring them together.

The balloon on the string will ascend to the ceiling, or at least move away, with surprising speed and vigor. Even in a large lecture hall, this demonstration will give every student a grandstand view.

ALSO: Charge a comb by combing your hair. Then bring the teeth of the comb near a thin stream of water from a faucet. The stream will be diverted and distorted noticeably and dramatically.

Determining Electroscope Voltage

When students have observed the operation of an electroscope, the teacher can ask one student to charge the instrument with his comb after he has combed his hair. This will cause the gold foil to stand at right angles to the vertical support. The teacher should ground out the electroscope with his finger in the usual way, and ask students what voltage they believe there was on the instrument to cause the effect they have observed. (Students' guesses usually range from microvolts to 10 volts.)

213

The teacher can then connect a penlight cell to the case and the conductor rod of the electroscope. Of course this action produces absolutely no observable effect. Next, he may connect a large (1½-volt) dry cell. (It possesses the same voltage as the penlight cell, but its size is impressive.) Then a 12-volt auto battery can be tried, with the same failing results. Finally, the teacher can bring out the variable D.C. power supply and volt meters, which at 600 volts causes the gold foil to rise to about 45°. At this point, students can be "warned" to be careful in the future about combing their hair; they may electrocute themselves!

NOTE: If the electroscope has calibrations, the students may be asked to plot a curve and extrapolate to determine how many volts are needed to make the foil rise to a right-angle position.

Using Homemade "Electroscopes"

After a series of teacher demonstrations of static electricity, the students may be given homemade "electroscopes" to perform their own individual demonstrations. These consist of a strip of newspaper 3 to 6 cm in width and about 30 cm long (60 cm long, folded in half). There is nothing critical about either the length or width of the paper.

To use his "electroscope," the student:

(a) Places the double strip in a textbook with several sheets of text between the two folds, leaving the end with the fold sticking out of the book.

(b) Pulls the exposed end smartly.

The rubbing friction of the sheets and strips will transfer enough charges so that the two strips will repel each other vigorously like an inverted "V", sometimes attaining angles up to 45°. Placing a hand between the two leaves causes them to "snap" at the hand. The leaves do not discharge too quickly.

HOMEMADE EQUIPMENT FOR STATIC ELECTRICITY

by J. R. Stevenson
Menlo Community School, Menlo, Iowa

Very often, when it comes time for the unit on static electricity, the teacher will find the school's supply of material to be limited to a vulcanite rod, a glass rod, a catskin, and a silk pad. The result: A few teacher or student demonstrations, with the rest of the class watching.

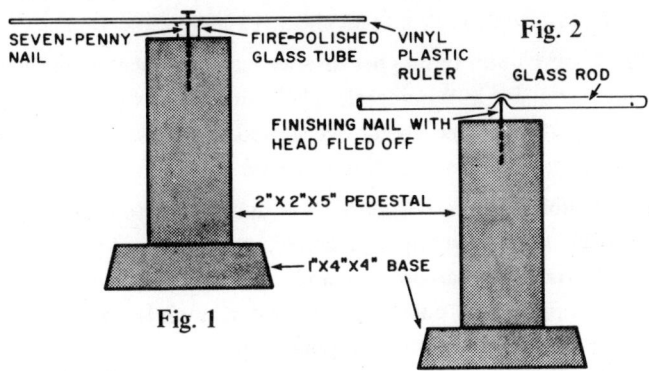

It doesn't have to be that way. With a modest expenditure of time, some scrap items from the school shop and lab, and a dollar or two to spend for miscellaneous items, a class can be equipped to investigate the principles of static electricity in depth. Here are three devices I have made for this purpose.

1. *Negative Charges:* An ordinary vinyl plastic ruler, when rubbed with a wool pad, can be charged negatively. You can supply each student with two such rulers, one of them mounted on a wooden base with a nail as a pivot and a glass tube as a bearing, as shown in Fig. 1. If one end of the mounted ruler is charged, it will be repelled when the similarly charged end of another plastic rule is brought close to it.

Building the device is simple: I use a plastic ruler of the type found in many five-and-tens or stationery stores with a hole already formed in the center (along with other slots and designs); a seven-penny nail fits through the hole very nicely.

> NOTE: **If you cannot locate this type of ruler, a solid ruler could be used. Form the hole by heating a nail red hot and thrusting it through the midpoint of the plastic ruler.**

I select a piece of glass tubing with a diameter that permits the nail to turn freely in it. I file about ¾ of an inch off the tubing and smooth each end with a file. To flare one end, I heat the piece of tube red hot, insert a tapered, round file into it as far as it will go, and press it against the heated glass tube. After the glass has cooled, I insert the nail through the center hole in the ruler and through the glass bearing, and pound the nail into the wooden stand, leaving just enough of the nail sticking out to permit the ruler to rotate freely.

> NOTE: **Flaring the end of the bearing is probably not necessary. Fire polishing it should be good enough to permit free rotation.**

2. *Positive Charges:* Following the same general principles, glass tubing from the chemistry lab can be used to make hollow glass rods that can be

positively charged when rubbed with silk. Here is how to make and mount the rods: Select a piece of tubing that is about ½ inch (15 mm) in diameter and 2 or 3 feet long (the finished rod should be about 12 inches long).

> NOTE: I use 15-mm tubing because we have it available, and because it is the same size as commercial glass rods. If, however, you do not have this particular size, the device should work just as well with any size rod from ¼ inches (7 or 8 mm) on up.

Heat the tubing; when it gets red hot and begins to soften, pull each end rapidly apart as in the making of chemistry pipettes. After the tubes have cooled, file off the long narrow tips and reheat to seal the pointed ends. The other (or open) end can be cut to the desired 12-inch length with a file or glass cutter and smoothed by filing and fire polishing.

Mount the rods on a base similar to that used for the plastic rulers, as follows: Determine the center of gravity of the tube and heat it at that point until it becomes red hot. While still red hot, remove the tube from the burner and bring it down over the pointed end of a finishing nail. This will force the softened bottom of the wall of the tube against the upper wall; additional pressure will cause a "dimple" to be raised in the upper wall that will serve as a bearing for the mounted tube, while the nail (with the head filed off) serves as a pivot (Fig. 2).

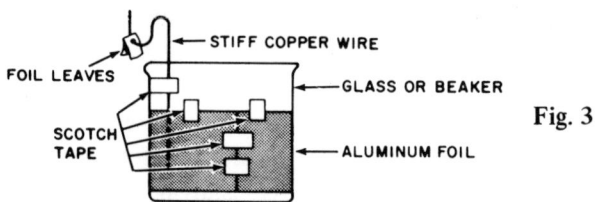

Fig. 3

In addition to using the glass rods to show the repulsion of like charges, you can use them in conjunction with the plastic rulers to show the attraction of unlike charges to each other.

COULOMB'S LAW: The glass rods and plastic rulers can be used in qualitative investigations of Coulomb's Law.

3. *A Simple Electroscope:* You can construct simple electroscopes in mass quantities very easily as shown in Fig. 3. Here are the steps:

(a) Wrap aluminum foil around a glass or beaker. Use a piece slightly bigger than the circumference of the beaker, and Scotch tape it together where it overlaps. At intervals, hold in place with Scotch tape around the top. This aluminum foil takes the place of the contact knob of the conventional electroscope.

(b) Bend a piece of stiff copper wire as shown in the illustration (Fig. 3) and insert between the aluminum foil and the beaker and secure with Scotch tape or masking tape.

(c) Cut leaves from aluminum foil and force over the hooked end of the copper wire.

In using this device, charged objects are presented to the electroscope via the foil wrapper around the beaker. The wrapper adds sufficient area to the electroscope to act as a "reservoir" for electrons that are forced into the leaves by the approach of a negative charge, or attracted (drawn) from the leaves by the approach of a positive charge.

A QUICKIE WITH THE ELECTROSCOPE

by George F. Smith

South Hadley High School, South Hadley, Massachusetts

Here is an interesting sideline demonstration to use when the electroscope is out for static electricity demonstrations. The electroscope can be used to introduce, qualitatively, the laws for series and parallel resistances. In the demonstration, I use fresh masking tape; it makes a good high resistance and is easily connected between the charging plate at the top of the electroscope and the metal shielding of the case. I usually use the same electroscope charged to the same amount for all three experiments. A variation, if there are enough electroscopes, would be to use three of them, all charged identically, and watch the three experiments below simultaneously.

First, I have the students count in unison to see how long the electroscope takes to discharge through a single 2-inch strip of masking tape (Fig. 1). Then I ask them to predict how long they think it will take to discharge the electroscope through two 2-inch strips in parallel (Fig. 2) and a single 4-inch strip (Fig. 3). Normally I find their intuition to be right, even though they have never studied current electricity. The application of so-called "common sense" is good science.

Typical Data

Single 2-inch strip	Two 2-inch strips in parallel	Single 4-inch strip
156 counts to discharge	81 counts to discharge	318 counts to discharge

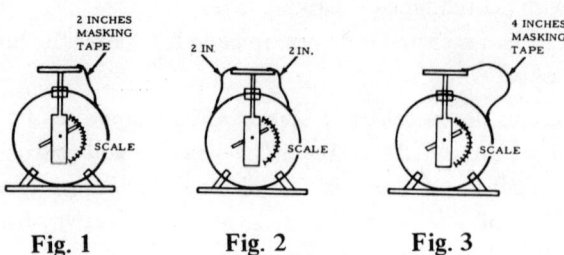

Fig. 1   Fig. 2   Fig. 3

12

CHARGES IN MOTION

TEACHER DEMONSTRATIONS IN ELECTRICITY, ELECTRONS, AND GASEOUS IONS

by Dudley W. Davis

Monroe Junior High School, Tampa, Florida

Here are a number of teacher demonstrations which I have found effective in creating student interest in electricity, electrons, and gaseous ions. Included also are directions for making an inexpensive Crookes tube. Each of the demonstrations can be carried out with a very small amount of science equipment, using items common to most homes or available at minimal cost.

> CAUTION: I have permitted students averaging 14 years of age to conduct all of these demonstrations under my supervision. However, all of the demonstrations here should first be performed by the teacher, and only then possibly conducted by selected students under the close supervision of the teacher.

Demonstrating High-Voltage Energy

Jacob's Ladder is our name for a demonstration apparatus which can be used to show students the energy produced by a high-voltage current, such as that produced by a spark plug in engines or that present in television sets and high-voltage power lines.

Materials: Materials needed for making a Jacob's Ladder are:

1 low-voltage power unit (or a 6- to 12-volt automobile storage battery)
1 induction coil
2 unpainted wire coat hangers
Pair of pliers (or wire cutters)
4 pieces of insulated wire, 16 to 22 gauge, 12" to 18" long

Procedure:

(1) Cut each of the coat hangers at one end of the bottom to obtain the longest possible straight piece. Make the other cut on each hanger approximately 4 inches around the corner at the other end. Then straighten the short end so that this piece is at an approximate right angle to the longer piece.

(2) Fasten the short ends of the cut-out hanger wire to each of the high-voltage binding posts of the induction coil, with the long ends of the wires in a vertical position (Fig. 1). Start with the nearest point between the wires at the bottom at about ½ inch or less, and shape the wires so that the tops are approximately ¾ inch apart, as indicated in the illustration.

(3) Attach the low-voltage power unit to the induction coil and turn it to *ON* or *INCREASE,* as appropriate. The arcs should jump between the two wires at the bottom, climb up the wires to the top ends, and make a popping sound as they jump off the tops. (Some minor adjustments of distances between the wires may have to be made to suit the output of the induction cell.)

WARNING: BE SURE TO TURN OFF OR DISCONNECT YOUR POWER UNIT BEFORE APPROACHING WITHIN 2 INCHES OF EITHER THE BINDING POST OR THE INDUCTION CELL. If 6 to 6.5 volts D.C. is used, the shock from the induction coil will not cause serious damage to a healthy person, but it is extremely uncomfortable.

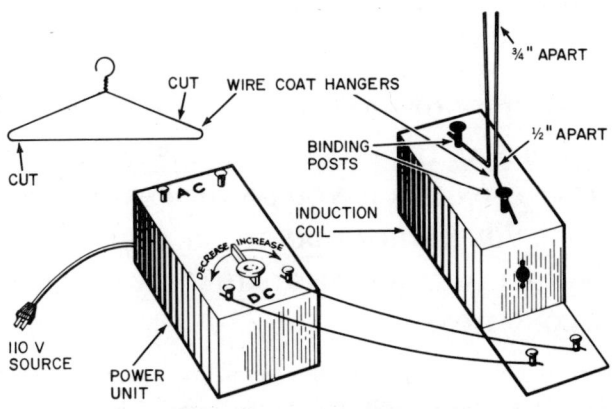

Fig. 1: Jacob's Ladder.

Demonstrating High-Voltage Penetration of Insulators

The following series of demonstrations shows students how high-voltage current penetrates some insulators to the point that they fail to insulate. Students should be able to apply their observations to various situations in their environment.

Materials: In addition to the basic apparatus used in the preceding demonstration procedure, you will need:

2 electrodes with insulated handles (attachments which usually come with induction coils)

NOTE: If the electrodes are not available, use two screwdrivers which are small enough to fit the binding post of the induction cell.

1 sheet of notebook paper
1 toy balloon about 3" long
1 common carbon lead pencil with eraser
1 used automobile condenser (capacitor)

NOTE: A capacitor can usually be obtained free for the asking from a garage.

Procedure:

(1) Turn off the power unit and remove the hanger wire used in the preceding demonstration from the induction coil.

(2) Insert the electrodes or screwdrivers through the high-voltage post binders. Hold the ends of the insulated handles and swing the metal tips toward each other to show how the arc travels from one to the other. Then arrange the electrodes so that one is at a right angle to the other and approximately ½ inch apart.

(3) Pass the piece of paper between the electrodes and watch the arc pass through the paper. If the paper is passed quickly, it will not ignite. If it is held still, it will ignite.

WARNING: DO NOT APPROACH AS NEAR EITHER OF THE ELECTRODES AS THE ELECTRODES ARE APART WHEN THE CURRENT IS ON, FOR IF YOU DO THE CURRENT WILL ARC THROUGH YOU RATHER THAN TO THE OTHER ELECTRODE. (18,000 VOLTS WILL ARC APPROXIMATELY 1" UNDER STANDARD CONDITIONS.)

(4) Turn the current off, and slip the balloon over one of the electrodes, being careful not to puncture it. (You may want to have one of the students inflate the balloon first.)

(5) When the balloon is in place, turn on the current and swing the electrodes so that they are near each other where the balloon is covering the one, and the arc will penetrate the balloon in one or more places as you move the uncovered electrode back and forth.

(6) Turn the current off, remove one of the electrodes, and insert the pointed end of the pencil into the high-voltage post binder.

(7) Turn on the current. Then with the voltage control switch on the power unit, turn the voltage down to almost off. Let one of the students who thinks that the pencil will not conduct electricity touch the eraser while you gradually turn up the voltage. The student will receive a small shock.

(8) Now swing the pencil and the electrode near each other so that the point of the electrode is at a right angle with the pencil and within ½ inch or

less. The electricity will penetrate the wood of the pencil at one or more places as you move the electrode back and forth along the pencil.

(9) Turn off the current, then attach the pigtail wire of the condenser to one of the post binders, and the body of the condenser to the other binder.

(10) Turn the voltage to high and leave it on for one to two seconds.

(11) Remove the condenser carefully, as it is now charged with the same amount of voltage as the induction coil puts out. It will arc the same as the electrodes. (A flashlight or automobile lamp bulb can be burnt out with the charge on the condenser.)

Interpreting effects: The pencil, balloon, and paper demonstrations show that voltage can be increased high enough so that some things we call insulators fail to insulate. This situation can be related by the teacher to high voltage such as that in the power lines of utility companies.

NOTE: Students should be warned never to get near one of these lines or attempt to move one that is down.

The condenser (capacitor) demonstration illustrates the danger of working on a television set or other electrical appliance, unless it is known how to safely bleed off the charge on the condensers. It takes up to 24 hours for some condensers to bleed off the charge by themselves.

A Crookes Tube to Demonstrate Electrical Discharge in Gases at Low Pressure

Materials: If you do not have a Crookes Tube or Cathode Ray Tube available to demonstrate gaseous ions, you can easily make one that operates quite effectively. You will need these materials:

1 glass tube, approximately 9" long and ¾" in diameter
1 rubber stopper without holes to fit the glass tube
1 one-hole stopper to fit the other end of the tube
1 small glass tube, approximately 3" long to fit the one-hole stopper
2 common four to six penny nails with medium or large heads.
1 electric vacuum pump

NOTE: Be certain to fire polish the glass tubing so that it will not crack under the reduced pressure.

Making the tube: Following are directions for assembling the tube:

(a) Drive one of the nails through the rubber stopper (without holes), from the small end of the stopper to the large end.

(b) Drive the other nail through the one-hole stopper in the same manner, being careful not to damage the hole.

(c) Insert the small glass tube through the hole in the rubber stopper so that it extends about ¼ inch through the small end of the stopper.

(d) Insert both stoppers tightly into the ends of the large glass tube. (A burette clamp and ring stand can be used to hold the tube. If these are unavailable, a small shoe box may be used: Stand the box on its end, cut a small hole through each side, and insert the glass tube through the holes.)

(e) Attach the wires from the induction coil to the points of the nails.

(f) Hook the vacuum pump to the small glass tube, turn on the power unit, and observe the Crookes Tube (Fig. 2).

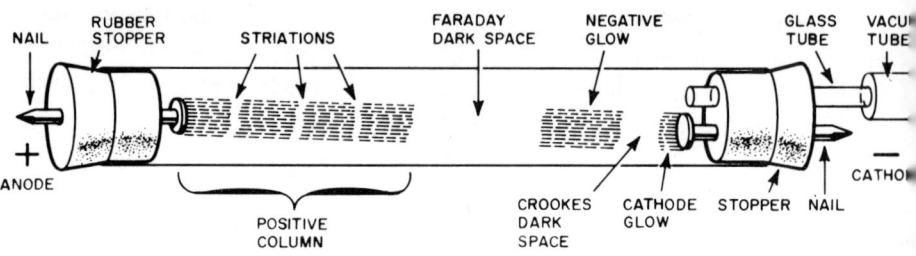

Fig. 2: Crookes Tube.

Using the tube: When an electrical discharge occurs between the two nail heads, some beautiful and interesting effects take place as the air is pumped out of the tube. When the pressure is reduced enough, a narrow pink streamer develops between the electrodes. As the pressure is further reduced, the streamer gets larger until it almost fills the tube.

If you continue reducing the pressure (below 1 mm of Hg), a state of discharge will be reached. The negative electrode (cathode) will develop a velvety glow known as the "cathode glow" (Fig. 2). Adjacent to this glow, toward the positive end, is an area with no streamer or glow that is known as the "Crookes dark space." Next comes the area that also glows, called the "negative glow." The second dark region follows this and is known as the "Faraday dark space." Last is the largest glowing area that reaches to the positive electrode (anode). This area is known as the "positive column."

The positive column is not a solid column, but is striated across as the electrons reach their energy levels. An ion must have a chance to travel far enough to acquire energy to produce ionization by collision. This can be accomplished by reducing the pressure in the tube. If one continues to reduce the pressure, the ions may go all the way to the other electrode without having a collision. In this case, the ions may eject electrons and other ions when they strike the electrode. These are then called "secondary ions" and are instrumental in supporting the discharge.

A bar magnet may be used to demonstrate the electrical charge of the particles. If it is placed so that the magnetic field is perpendicular to the component of motion of the charge, the stream will bend in one direction perpendicular to both the field and the original velocity of the charge. The stream will bend in the opposite direction, again perpendicular to both the original velocity and the magnetic field if the charge is opposite or if the charge is the same but the magnetic field is opposite.

> NOTE: The rays are deflected by the electrostatic field. By applying a potential difference between plates on opposite sides of the tube, the positive plate attracts and the negative plate repels the stream.

If the pressure in the tube is further reduced, the positive column shortens and the Faraday dark space lengthens. This is followed by a greenish glow at the glass. If reduction of pressure continues, this greenish glow will spread over almost the entire area of the tube. It is due to the fluorescence of the glass caused by the bombardment of rays from the negative electrode of the tube. These particles are called "cathode rays."

PRACTICAL REACTANCE MEASUREMENTS

by Lawrence B. Ryan

Beaver River Central School, Beaver Falls, New York

Most physics laboratory manuals include some exercise designed to show, usually rather qualitatively, that there is a difference in behavior to coils from A.C. and D.C. The exercise generally involves winding a coil and noting its effect on the brilliance of a series lamp. This method has a serious disadvantage, however: It ignores the vital frequency dependence of reactance, simply because line supplies maintain an almost unvarying frequency of 60 Hertz.

> NOTE: The following exercise is designed to rectify this neglect of reactance in the traditional exercise, using a source of *variable frequency* alternating current, the common laboratory audio generator.

Materials

Equipment needed for the experiment includes:

Audio generator
6-volt lantern battery
0 -6 D.C. voltmeter
0 -1 D.C. ammeter
0 -15 A.C. voltmeter
0 -10 A.C. *milliammeter*

Coil, iron core: D.C. resistance∿60Ω.(The primary winding of a small 110, 6-volt, step-down transformer is excellent.)

Condenser: 0.5 - 2uf, nonelectrolytic (Oil-filled surplus units are ideal.)

NOTE: To be useful, the audio generator must be capable of an output of at least 10 -15 volts at about 10 milliamps. Units from Heath, Eico, etc., are good.

Procedure

(1) Connect the circuit of Fig. 1.

(2) Using the voltmeter-ammeter method, measure the resistance of the coil. (The value obtained is the pure D.C., or "ohmic," resistance, R.)

(3) Connect the circuit of Fig. 2.

(4) Set the generator's frequency to 60 Hz (line frequency).

ACCURACY: If very precise work is desired, the frequency of the generator can be checked at each step by beating against line frequency, using Lissajous figures on an oscilloscope.

(5) Starting at zero, advance the generator's voltage (output, or gain) control, taking readings of the current through the coil at 7 or 8 steps over the available range.

(6) Plot the data. (The result will be a straight line whose slope, from Ohm's Law, is the "resistance" of the coil on 60 Hz A.C.)

GRAPHING: In plotting the graph, the normal axes must be "reversed." Current must be plotted horizontally and voltage vertically so that the slope will have the units of "ohms."

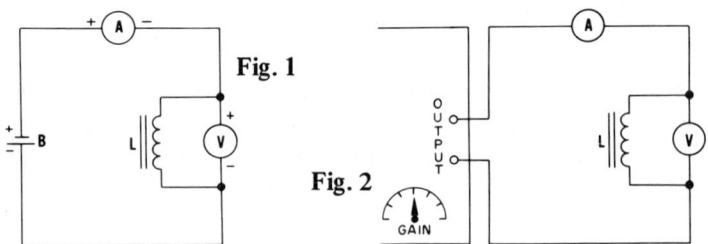

The graph will point out two things:

(a)   Since it is a straight line, the "resistance" is a constant value at constant frequency;

(b)   This value is considerably higher than the "ohmic" resistance from step 2.

This resistance is, of course, the impedance, Z, of the coil at 60 Hz.

(7) Now set the generator's output voltage to about 10 volts. Holding this

value constant, read the current through the coil as the frequency is varied in 20-Hz steps from 20 to 200 Hz.

> NOTE: Here is where the frequency dependence of reactance becomes vividly clear, for, as the frequency of the applied emf increases, the current in the coil *decreases* with no change in emf. This can only come about through an *increase* in the "resistance" (impedance) of the coil.

Using several of the data points from this section, and with the appropriate formulae, the student can calculate the impedance, Z, and the reactance, X_1, at each point. From these, the inductance, L, of the coil can be figured and this will be seen to be a constant quality of the coil.

(8) Apply the same technique to the condenser. (Of course the condenser will show no current on D.C., indicating "infinite" resistance. On A.C., the condenser will "pass" a current and will, like the coil, have a constant "resistance" at a constant frequency.)

(9) If step 7 is repeated with the condenser, it is found that as the frequency increases, the current through the condenser *also* increases, indicating that its "resistance" (reactance) *decreases* with increasing frequency.

(10) If the values of parts are carefully chosen, the circuit of Fig. 3 can be

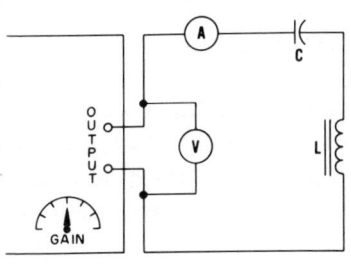

Fig. 3

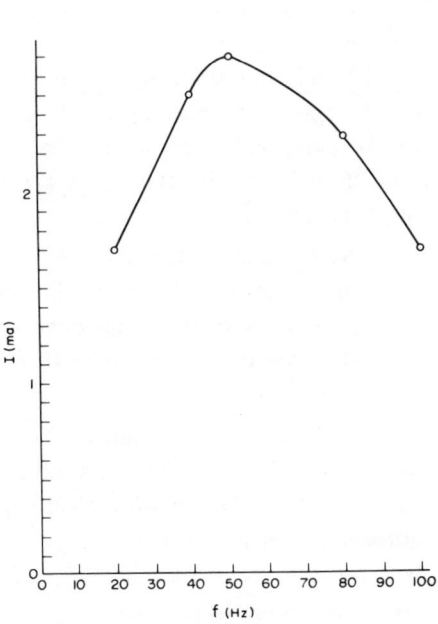

Series Resonance Plot

Fig. 4

connected. Keeping the voltage constant, the frequency is varied from 20-200 Hz. At some point in this range, as shown in Fig. 4, the current will rise to a maximum. This point is, of course, the series resonance frequency, and its value will compare nicely with the theoretical value from the equation

$$f = \frac{1}{2\pi\sqrt{LC}}$$.

PRODUCTION OF MOTION FROM ELECTRICITY

by Ralph S. Vrana

California State Polytechnic College, San Luis Obispo, California

Here are a series of experiments which demonstrate the production of motion from electricity.

1. **Rotating bar magnet:** The principle of how an electric motor operates can be shown by placing a bar magnet on a compass stand. Two other bar magnets are held as shown in Fig. 1 to start the rotation (the hand magnets must be switched simultaneously and at the proper moment to keep the pivoted magnet rotating).

> NOTE: This is the principle of how an electric motor operates. In many motors, of course, the rotating magnet (armature) shifts its polarity by switching the direction of current in its windings. In such cases there is no need to shift the direction of the hand-held magnets (field coils).

2. **Demonstration motor:** You can make a satisfactory demonstration motor from the armature of a small electric motor, such as is found in a vacuum cleaner, auto heater, or windshield wiper. Disassemble the motor to remove the armature, which will come out in one piece, then mount the armature on a board so that the shaft is held at each end by a screw-eye and is free to rotate easily. A strong, heavy magnet is placed near the middle of the armature and wires from the power supply or storage battery are held against the commutator of the armature (Fig. 2). These wires are the "brushes" of a conventional electric motor.

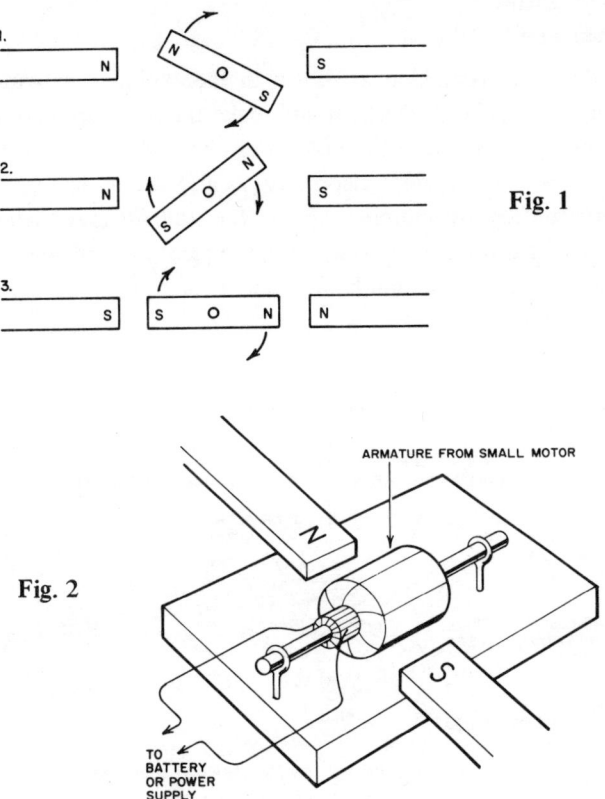

Fig. 1

Fig. 2

ARMATURE FROM SMALL MOTOR

TO
BATTERY
OR POWER
SUPPLY

NOTE: A little adjustment may be necessary before the armature will spin. The magnet works best when held very close to the armature, but may make contact if brought too close. Several good bar magnets placed close to each side of the armature will also get the motor going. The bar magnets on one side must have opposite polarity to those on the other.

This rejuvenation of an electric motor is a good learning experience, because the essential parts of the motor are out in the open, yet the motor runs nearly as rapidly as it did in the original, closed housing. The speed is quite responsive to the movement of the field magnets and to pressure and position of the brushes.

3. Making an electric motor: To make a homemade motor commutator and armature, you will need:

Shaft from coat hanger

Adhesive tape

Tin can (for strips of metal)

Two long nails

Copper wire

The device is assembled as shown in the exploded drawing of Fig. 3. In building the motor, the students should make the armature first, and hang it on bearings (Fig. 4). This much of the motor can be tested with permanent magnets and a source of direct current electricity, as also shown in Fig. 4. Then, if the motor passes the test by spinning rapidly, the students may then go on to make the field coils (electromagnets which take the place of permanent magnets). If it doesn't spin, the trouble should be found and corrected. Here are precautions to take in construction:

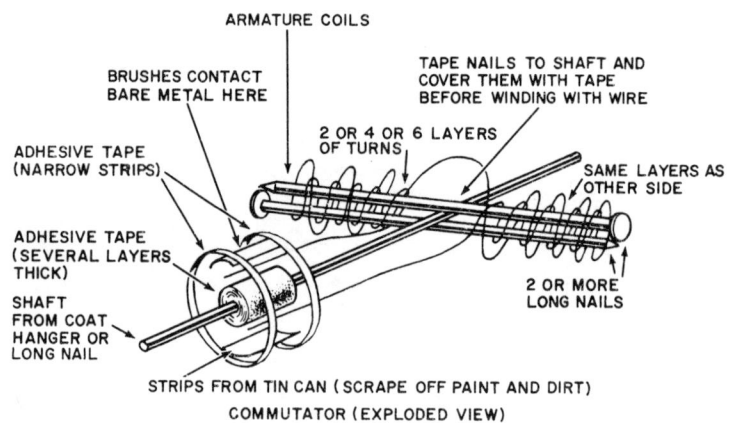

HOMEMADE MOTOR
COMMUTATOR AND ARMATURE

Fig. 3

(a) Be sure the copper wire is cleaned of insulation (including enamel) wherever it is to be connected.

(b) Tin can material may be covered with an enameled surface of the same color as the tin. This must be scraped off with a piece of sandpaper or file wherever an electrical connection is to be made.

(c) When winding copper wire onto a strip of tin can, it is a good idea to cover the tin can material with one or two layers of adhesive or masking tape in order to prevent a short circuit between tin can and wire.

(d) If the electric motor contains more than one "field coil," these must be connected so as to produce a magnetic field which has a south pole on one side of the armature and a north pole on the other (Fig. 5).

(e) The commutator (contact points between the ends of the armature coil and the brushes) must be adjusted for maximum speed by twisting it on its housing.

(f) Brushes must touch the commutator slightly.

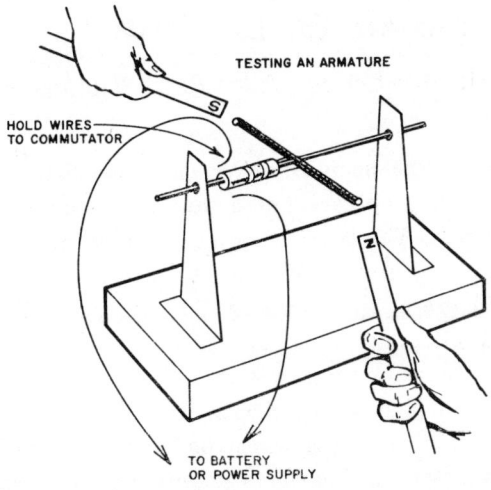

Fig. 4

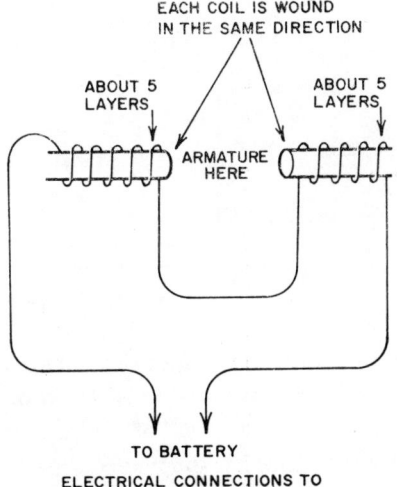

Fig. 5

DEMONSTRATING CAPACITOR
ACTION AND ELECTRON FLOW

by David D. Lockhart

Allen Park High School, Allen Park, Michigan

The simple circuit of the neon lamp relaxation oscillator provides an excellent means of demonstrating capacitor action and electron flow to the beginning electronics student. The circuit depends upon a unique property of neon lamps—an ignition voltage somewhat higher than their extinction voltage. A typical Ne-2 lamp will fire at approximately 70 v.d.c. and will continue to conduct until the applied voltage falls to about 50 v.d.c.

As shown in Fig. 1, the circuit is easy to bread-board and its components are readily available in most high school physics labs. The part values are not critical and different capacitors and resistors can be inserted into the circuit and their effect on frequency observed. The values suggested in the diagram (Fig. 1) will produce pulses of approximately 16 Hz.

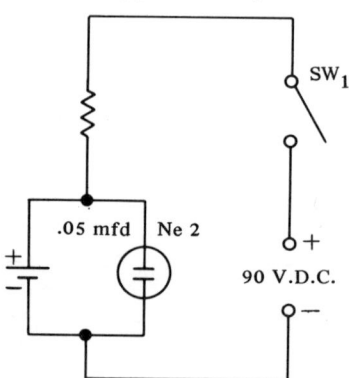

Fig. 1: Relaxation
Oscillator

Circuit Operation

The dynamic operation of the circuit can be described as follows:

At the instant the switch is closed on the relaxation oscillator, electrons flow toward the negative side of the capacitor and, simultaneously, away from the positive side. This has the immediate effect of producing a large voltage drop across the resistor.

As the charge across the capacitor increases, the electron flow diminishes, and simultaneously, the voltage drop across the resistor decreases. As the charge across the capacitor rises to the ignition voltage of the Ne-2 lamp, it fires.

Subsequently, the electrons which collected on the capacitor producing the charge move through the lamp, thus discharging the capacitor. As the voltage across the capacitor falls below the extinction voltage of the lamp, it ceases to conduct. Once again, the voltage begins to rise across the capacitor as a proportional drop appears across the series resistor. In this manner, the cycle repeats itself.

Applications

After students have observed capacitor action and electron flow across the circuit, they can be shown the effects of using either a larger capacitor or a larger series resistor. A larger capacitor requires more time to charge and the frequency is slowed. A larger series resistor reduces the rate at which electrons can move away from the positive side of the capacitor and will also decrease the frequency.

> **NOTE: An oscilloscope offers a convenient means of illustrating the dynamic voltage changes in the circuit. The voltage variations are characteristic of changing capacitors and provide an introduction to a discussion of time constants (Fig. 2).**

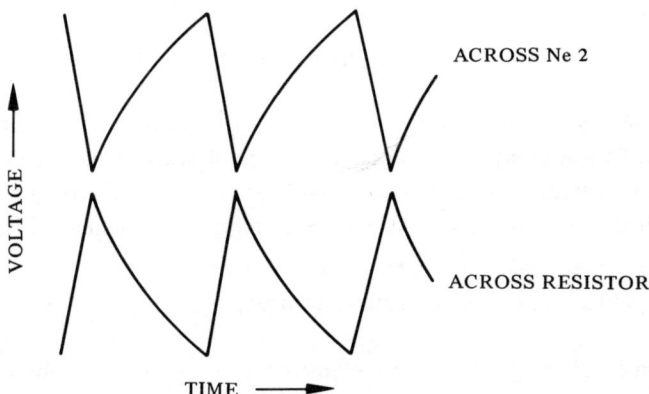

Fig. 2: Voltage across circuit components.

MEASURING THE MAGNITUDE OF THE ELEMENTARY CHARGE IN COULOMBS

by Don L. Hein

Southwest Minnesota State College, Marshall, Minnesota

Since many physics students find the concept of the elementary charge vague and meaningless without a "concrete number" to associate with the charge, I use the following experiment to give them the needed reference. This simple experiment, which makes use of the electrochemical approaches of PSSC to electric current, yields a measurement of the magnitude of the elementary charge in coulombs, perhaps the most common unit for expressing the quantity of electrical charge.

Problem

If we could measure the total quantity of a sample charge (in coulombs) transported from one place to another and simultaneously count the number of elementary charges that carried the sample charge, we could then determine the magnitude of the elementary charge in coulombs. In mathematical form:

If Q = Total quantity of charge in coulombs
N = Total number of elementary charges,

then $e = \dfrac{Q}{N}$ (Magnitude of the elementary charge in coulombs.)

Explanation

Copper sulfate ($CuSO_4$) is a crystal that consists of copper, sulfur, and oxygen atoms in a chemical arrangement. When copper sulfate is dissolved in water, the molecules dissociate into a copper ion (Cu^{++}) with two positive elementary charges and a sulfate ion (SO_4^{--}) with two negative elementary charges. These ions are more or less free ions when in solution.

Metallic copper is also a crystalline substance made up of copper atoms (assume to be pure). Metallic copper, however, is a good conductor of electric current, which implies that the negative elementary charges of the copper atoms

234

must be loosely bound and are free to drift through the crystalline structure. This also implies that copper atoms, in the metallic structure, can also become ions which can carry either one or two positive elementary charges.

NOTE: For greater detail, the instructor can refer to more complete explanations of the preceding.

Experimental Procedure

To conduct the experiment:

1. Set up an electrolytic cell using a strip of copper and a carbon rod, as shown in Fig. 1.
2. Suspend the two electrodes in a saturated solution of copper sulfate.
3. Connect the copper strip to the positive terminal of a direct current source and the carbon rod to the negative terminal.

Results

The free negative elementary charges will be removed from the carbon strip, making it positive. At the same time, elementary charges will be forced on the carbon rod, making it negative. The ions of copper (Cu^{++}) in solution will then drift toward the carbon rod where they will be neutralized and deposited. The sulfate ions (SO_4^{--}) in solution will drift toward the copper strip and be neutralized also, but in being neutralized they will remove one Cu^{++} from the strip and form $CuSO_4$. This will then go back into solution, thus replacing the Cu^{++} which was removed at the carbon electrode.

To summarize:

(1) At the positive terminal, copper electrode, Cu^{++} ions are put into solution.
(2) At the negative terminal, carbon electrode, Cu^{++} ions are taken out of solution.
(3) The over-all effect is that for every Cu^{++} taken out of solution, two negative elementary charges are used.

Measurement

What must be measured is: (1) total quantity of charge that has been used, and (2) the number of elementary charges used. The first calculation can be made by measuring the electric current and the time it was used:

$$\text{Quantity of Charge} = \text{Current} \times \text{Time} , \text{ or } Q = IT.$$
$$\text{(coulombs)} \qquad \text{(amperes) (seconds)}$$

Measurement of the number. of charges used can be made by using the definition of a mole, Avogadro's Number, and the total mass of copper deposited on the carbon electrode.

> **NOTE: The Cu-plated carbon electrode should be dried before reweighing, but** *do not wipe it dry.* **Shaking off droplets and baking the electrode under an incandescent bulb for a short time should be sufficient.**

If M_{cu} is defined as the mass of copper deposited (M_{cu} = Final Mass of Carbon Rod - Initial Mass of Carbon Rod) and the atomic mass of copper is taken to be 63.5 amu, then the mass of 1 mole of copper is 63.5 g, which contains 6.023×10^{23} atoms according to Avogadro's Hypothesis. The number of elementary charges used (N) will be:

$$N = \frac{(M_{cu}) (6.023 \times 10^{23}) \text{ atom/mole}}{63.5 \text{g/mole}} \times 2 \text{ elementary.} \quad \text{charges/atom}$$

This equation can be derived as follows:

$$\frac{M_{cu}}{63.5 \text{ g/mole}} = \frac{X}{6.02 \times 10^3 \text{ atom/mole}}.$$

[Solve for X, which represents the number of copper atoms deposited on the carbon rod. Then each atom requires 2 elementary charges to become neutralized. The final value (e) is merely the division of total charge (Q) by the number of elementary charges(N).]

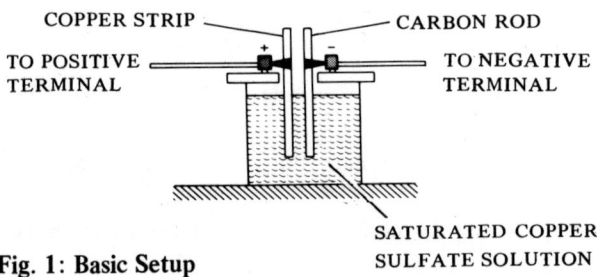

COPPER STRIP ⎯ ⎯ CARBON ROD

TO POSITIVE TERMINAL ⎯ ⎯ TO NEGATIVE TERMINAL

Fig. 1: Basic Setup

SATURATED COPPER SULFATE SOLUTION

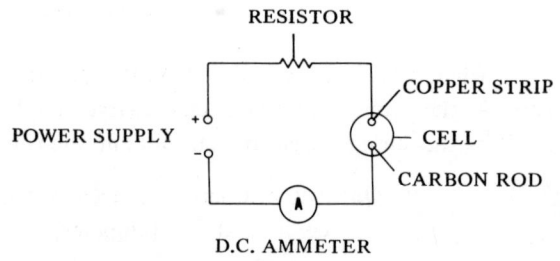

RESISTOR

COPPER STRIP

POWER SUPPLY

CELL

CARBON ROD

D.C. AMMETER

Fig. 2: Circuit Diagram

General Comments

To achieve good results, the surfaces of the carbon rod and copper strip must be clean. (Sandpaper might be used for cleaning.) The current in the cell should be about 0.5 amperes obtained from either a 110-V source D.C. used in series with a 60-W lamp, or a 6-V source connected in series with a 4-ohm resistor (Fig. 2). For best results the cell should be operated 30 minutes to 60 minutes.

ACCURACY: Do not expect results to be extremely accurate since there are several sources of error involved, but they should be in the same order of magnitude (10^{18} elementary charges).

13

USING PHOTOGRAPHY
IN THE LABORATORY

LISSAJOUS FIGURES:
A LABORATORY EXERCISE

by John H. Jeffers

Berkeley Preparatory School, Tampa, Florida

Following are background materials and an interesting method for making and recording the patterns traced by a double pendulum, called Lissajous figures. The latter offers the physics teacher an effective alternative to the usual demonstrations of these figures with an oscilloscope and electron beam, or a double pendulum rigged to release a fine trail of sand or salt as it moves.

Pendulum Patterns

A double pendulum is a mass suspended in such a manner that its frequencies in both the x and y directions are different. This setup is most easily constructed by suspending the mass from a three-string arrangement as diagrammed in Fig. 1. Fig. 1a shows the front view and Fig. 1b, the side view. The pendulum can swing only in the x direction, with the length of string labeled L_x. With the length L_y, the mass can swing only in the y direction.

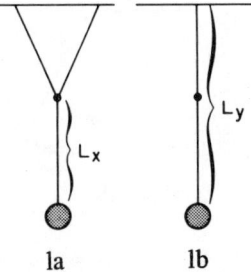

la   lb

Fig. 1

For a simple pendulum, when the amplitude of swing is small, the frequency is given by the expression:

$$f = \frac{1}{2\pi}\sqrt{\frac{g}{L}},$$

where

L is the length of the pendulum and g is the acceleration of gravity. When a

239

double pendulum is swung, the mass oscillates in both the x and y directions. The complex pattern of motion is called a "Lissajous figure."

With a double pendulum, the ratio of the frequencies can be calculated by the expression:

$$\frac{f_x}{f_y} = \frac{\tfrac{1}{2}\pi \ \sqrt{g/L_x}}{\tfrac{1}{2}\pi \ \sqrt{g/L_y}} = \sqrt{\frac{L_y}{L_x}} \ .$$

The ratio of the frequencies equals the square root of the inverse of the ratio of the lengths.

In the limiting case, the simple pendulum $L_x = L_y$; therefore, $L_y/L_x = 1$; $L_y/L_x=1$ and $f_x/f_y=1$. Thus, the frequencies in both directions are the same. A simple pendulum, if given momentum only in the x plane, will swing in a straight line in the x direction. If the pendulum is given equal amounts of momentum in both the x and y directions, the mass will describe a circle (Fig. 2a). If the pendulum is given unequal momentum, the path of the mass will be an ellipse (Fig. 2b). In any case, the pendulum will complete its swing in one direction just as it completes its motion in the other direction.

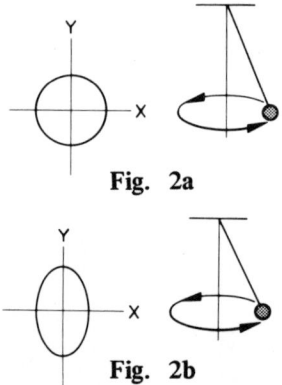

Fig. 2a

Fig. 2b

However, if the lengths L_x and L_y are not equal, the frequencies f_x and f_y will not be equal. Thus the mass will oscillate a different number of times in one direction than the other. When $f_x/f_y = 1:2$, the mass will traverse the y direction twice for each time it swings once in the x direction and will have a path as shown in Fig. 3.

SOURCE: An excellent summary of the patterns produced by various frequency ratios appears in *University Physics,* 3rd ed. (F. W. Sears and M. W. Zemansky, Addison-Wesley Pub. Co., Inc., 1964), page 270.

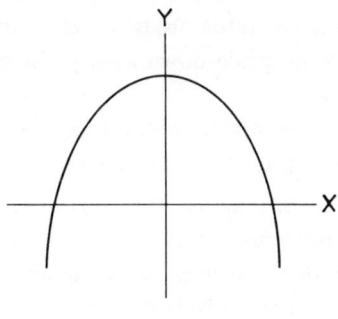

Fig. 3

Demonstrating Lissajous Figures

Various methods: Often Lissajous figures are demonstrated by applying different frequency signals to the vertical and horizontal inputs on an oscilloscope. The electron beam traces out the path of the figures on the screen. A drawback to this system, however, is that it obscures the relation of the figures to a pendulum. As noted, too, sometimes the figures are shown with a double pendulum constructed so as to release a fine trail of sand or salt as it moves.

As an alternative to these methods, my advanced placement physics class developed another method.[1] With this method, a can of aerosol shaving cream is used as the mass of the pendulum. The can is fixed to release a trail of shaving cream on a table top to produce the Lissajous figure, and the figure is then photographed.

Materials: Materials required to construct the apparatus involved include:

11 oz. can of Aero Shave

SOURCE: Aero Shave shaving cream is produced by Boyle–Midway Inc., Cranford, New Jersey 07016.

Tape
Wire hook
String
Plastic protective cover (to fit the base of the can) which comes with the
    can
Paper clip

Procedure: Directions for setting up this apparatus (Fig. 4) are as follows:

1. Place a plastic cover on the bottom of a can of Aero Shave.
2. Firmly tape a wire hook to the side of the can near the top.

---

[1] Actual construction of the pendulum evolved from an initial suggestion by the author. Many modifications were made by students R. Brown, A. Carron, R. Messner, and J. Widler.

3. Tie a piece of string to the spout, lead the string through the wire hook, and fasten it to the plastic cover on the bottom of the can.

4. Suspend the can upside down by a paper clip hook in the center of the plastic cover.

5. To keep the spout open, twist the cover to pull the string taut and create sufficient pressure on the spout.

Using the apparatus: In practice, the ratio of L_y/L_x is calculated for the desired f_x/f_y. To adjust the actual lengths, move a slip knot to the proper point on the upper two strands of the strings. Several practice swings will be necessary to get the feel for the amount of displacement from the equilibrium point and the amount of energy that should be imparted to the mass.

Hold a piece of paper under the can while twisting the cover to start the flow. Release the can, thus allowing it to produce the Lissajous figure. (Usually we allow the pendulum to swing through the full pattern and retrace it once.)

Taking photographs: To record the figures permanently, they may be photographed easily. We have used a Polaroid 210 with type 3000 film for this purpose with good results.

NOTE: The pictures should be developed about twice as long as the recommended time to increase contrast.

One photograph accompanying this report (Fig. 5) was taken with a Honeywell Pentex using Tri-X film, a flash gun, and a lens opening of f/16. Figs. 6 and 7 were taken with the Polaroid camera.

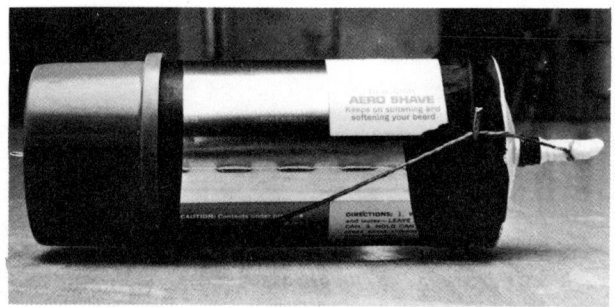

Fig. 4

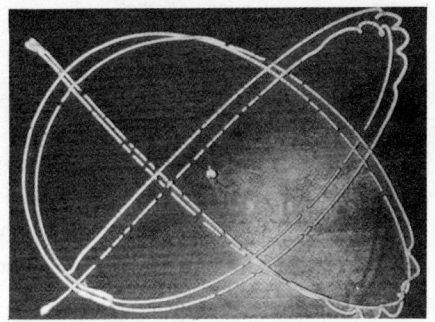

Fig. 5: $f_x/f_y = 4:5$

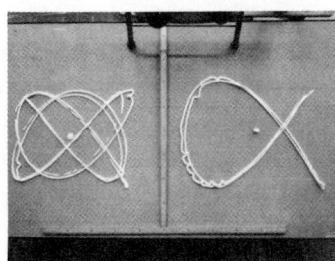

$W_x/W_y = 5:7$ $W_x/W_y = 2:3$

Fig. 6

$f_y/f_x = 3:4$ $f_y/f_x = 3:5$

Fig. 7

PHOTOGRAPHY IN
THE PHYSICS LABORATORY

by Ray D. Holland

Lawton High School, Lawton, Oklahoma

Using multiple flash photography in the physics laboratory to study kinematics and dynamics is certainly not novel. There are various ways of

243

exposing the film, making measurements from the film, and applying the results to teaching. Different types of moving objects under varying conditions can be used as study subjects.

However, the method and applications of photography described here offer the special advantage of moderate cost. The technique requires a camera with adjustable shutter opening and a time exposure mechanism, a glass table top, two laboratory-made air pucks, and a stroboscope, which can be purchased for about $70. (The adjustable shutter opening is not necessary, but helps in getting pictures of good quality.)

Making and Setting Up Materials

The air pucks required for this method are of the type described in PSSC literature. Each wooden disk is about 6 inches in diameter and 1 inch thick. A hole is drilled halfway through the disk at the center to receive a section of glass tubing, and a small hole the size of a large needle is continued through the disk (Fig. 1). Some experimenting with the size of the small hole, the air outlet, is necessary.

When the holes have been made, a piece of glass tubing about 1 1/2 inches in length is cemented in the top hole and a balloon is attached to the tubing by means of a #2 one-hole rubber stopper. The bottom of the disk should be sanded flat.

NOTE: For pictures to show the conservation of momentum, conditions are simplified if the two disks are of approximately the same mass, although disks of different mass could be used and the grid system could be used to interpret actual velocity and thus give the actual momentum.

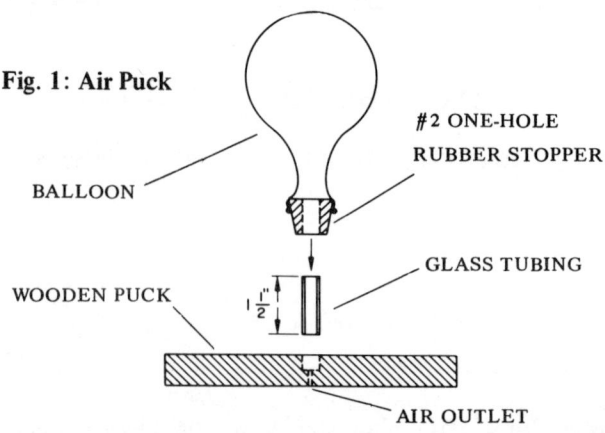

Fig. 1: Air Puck

#2 ONE-HOLE
RUBBER STOPPER

BALLOON

GLASS TUBING

WOODEN PUCK

AIR OUTLET

The air pucks operate best on a piece of flat glass plate resting on a table or other suitable area. (The plate shown in the accompanying photos is 80 cm by 150 cm.) For purposes of measuring displacements in the pictures, narrow strips of white masking tape are stuck to the bottom of the glass to form a grid of squares measuring 5 cm on each edge. Sheets of black construction paper are placed between the glass plate and the table to provide a black background.

Pucks are painted white on top. The balloons used are small, round, and dark in color. If the balloons are light colored, they interfere with the leading edge of the disk in the previous exposure.

Taking the Pictures

The group of students taking the pictures should work in a fairly dark room. To take the photos shown here, the camera was placed about 6 feet above the center of the glass plate and arranged to operate on time exposure. In this case, Triple-X Pan Kodak film was used and the shutter opening set at $f/11$.

Since the stroboscope has a parabolic reflector which produces a spot of light, the person operating the strobe had to follow the action of the disk with the spot for good exposure of the film. (The strobe rate used in these pictures was six flashes per second.) Another person operated the camera to keep the shutter open, and a third and fourth handled the disks.

> **PRACTICE: All of this requires a certain amount of practice and coordination. Our groups exposed five or six rolls of film before finding the right conditions for pictures which were satisfactory for calculation.**

Using the Pictures

We use the photos in various ways. The photography department prints a variety of large and small pictures for the bulletin board and a quantity of 3" x 5" prints similar to Photos 1 and 2 to give to students for a take-home assignment. Students are asked to find the velocity in Photo 1 and the acceleration in Photo 2. (In finding the acceleration, some students will assume a position for zero velocity and apply the formula $a = 2d/t^2$, while others will plot average velocity between successive exposures vs time and take the slope of the line as the acceleration. A few will hit upon the idea of taking the average velocity between successive exposures at different times during the trip and using the formula $a = \Delta v/\Delta t$.) In class discussion, negatives are used on the overhead projector for analysis.

Pictures of two-disk collisions are used to discuss conservation of momentum and kinetic energy. Though the regular-sized negative can be shown on the overhead projector, I have found that this particular analysis is more

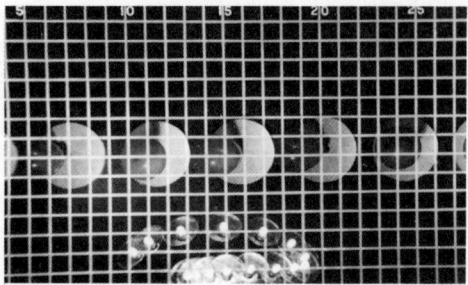

Photo 1: Disk moving at constant speed. The bright spots at the bottom of the photograph are reflections of the stroboscope from the top of the glass plate.

Photo 2: Disk accelerating down an incline.

effective if an enlarged 10'' x 6'' positive transparency is used. In practice, a sheet of acetate is placed over the transparency so vectors can be drawn without damaging it.

Since the disks have the same mass, we can argue for conservation of momentum if the vector displacement of the striking disk during one flash is equal to the vector sum of the displacements of the two disks during one flash after collision. The discussion for this is similar to that given in the *Teacher's Guide for PSSC Physics* for the experiment on collisions in two dimensions with .steel spheres.[1]

PHOTO 3: A glance at Photo 3 shows that kinetic energy is not conserved. Since the two masses are the same, the angle of separation between them would be 90 degrees if kinetic energy were conserved.

---

[1]Physical Science Study Committee, *Teacher's Resource and Guide Book—Part III* (Boston: D.C. Heath & Co., 1966), III-9 (1).

Photo 3: Collision between a moving and a stationary disk.

Variations

Other aspects of motion can also be explored using this method. Photo 4 was taken of two disks of different mass tied together with a white string. A small, white pith ball was located on the string at the center of the mass of the system, and the two disks were sent oscillating across the glass plate. The photo shows the center of the mass, where the pith ball is located, travels along a straight line. This exercise is usually performed during class with the Polaroid camera so that the results can be discussed immediately. It is interesting to compare the picture with one taken when the pith ball is not at the center of the mass.

NOTE: Although we do take some pictures during regular class time in a few cases, it is generally better to have a group of students come in during a free period to do this work. The pictures can then be used to augment the lab work done by the whole class with more basic laboratory apparatus.

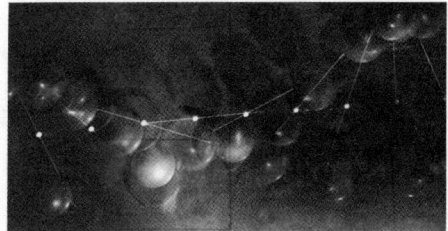

Photo 4: Picture showing motion of center of mass. The disks were painted black in this picture so they would not interfere with the image of the pith ball.

AN INEXPENSIVE SYSTEM FOR COLLISIONS IN TWO DIMENSIONS

by R. S. Dickens

Nathan B. Forrest Senior High School
Jacksonville, Florida

One of the most effective means of demonstrating the conservation of momentum within a mass system to high school students is with multiple flash photography. The object whose motion is to be observed is identified by a reflective mark while its surroundings are rendered flat black. As the object moves, it is periodically illuminated with a high-intensity strobe light or periodically observed through a moving slit. This is recorded on a single photographic plate, where the movement of the subject appears as a series of marks.

To conduct such an analysis, the only equipment required is a strobe light or strobe disc and an inexpensive camera. However, this demonstration is very difficult to perform without the elimination of friction forces in the interacting masses.

NOTE: For the photographs to be meaningful, it is necessary for the objects to move without friction. The purpose here is to describe a nearly frictionless system which is low enough in cost and simple enough in construction to be of use to the high school physics teacher who is limited in both resources and time.

Making a frictionless surface: To create this relatively frictionless system, first obtain some of the plastic substance sold as Dylite spheres, which takes the form of very small globules (about the size of a table salt crystal).[1] These spheres are remarkably round, and when sprinkled over a flat area act as a very good frictionless surface for almost any object that is flat. While there is more friction using the spheres than with a gas bearing, the distance photographed will rarely cover more than a meter, and within this distance the friction is negligible.

---

[1]One source of this material is Ayer Sales, Inc., 210 Highland Ave., Somerville, Mass. 02143. It is listed as Dylite spheres F-40C, and costs about $2 per pound, easily enough for a year's supply.

Next, construct a simple table to provide a large surface area, minimum glare, maximum flatness, and easy leveling. Plans for such a table are shown in Fig. 1.

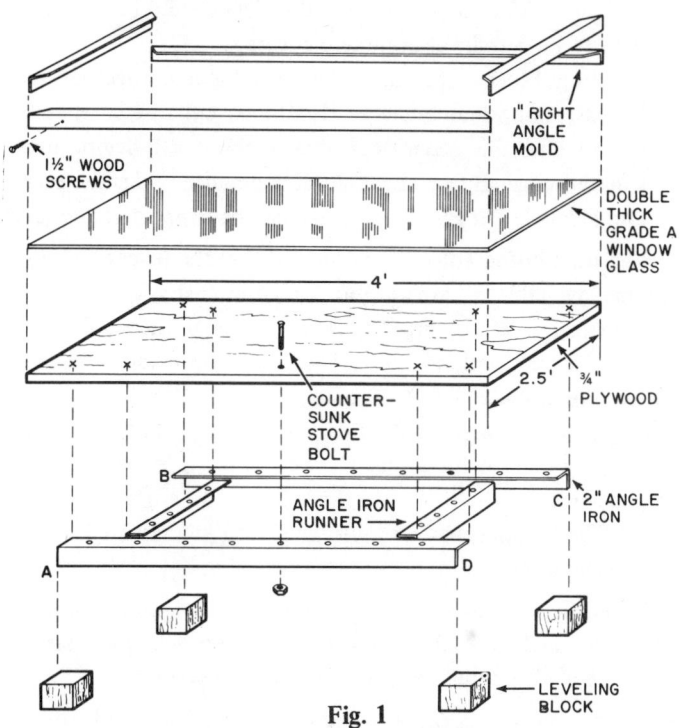

Fig. 1

(a) The plywood is ¾" thick and has an area dimension of 2.5' x 4'. Two-inch angle iron is used to improve the flatness of the table. This should be bolted to the plywood base with flat-headed stove bolts countersunk into the plywood and placed about 10 inches apart.

(b) Exact spacing of the angle iron runners is not critical, but the two runners parallel to the long edge should be about 3 inches inside of the edge. Note that the angle iron pieces are not connected to each other. They prevent the plywood from warping and give the table enough weight to press it down firmly on all of the supports. The most critical part of the table is the glass surface. Grade A, double-thickness window glass is sufficient for the surface and will keep costs down. To prevent glare from the glass plate, the rear surface of the plate should be painted flat black, using a high-quality spray paint.

NOTE: An acceptable reduction of the glare can be gotten only by painting the glass itself. **The placing of a black background under the glass will not be sufficient.**

It is important that the glass surface be extremely clean before painting. Three coats will probably be necessary to completely cover the glass. After the final coat, the glass should be held up to a bright light to be certain of complete coverage.

(c) Leveling blocks can be made from gluing together two 2" x 4" boards. Four 6-inch pieces should be cut out so that one edge is square and the other edge makes about a 5-degree incline to the squared edge. By placing the inclined edge under the points A, B, C, and D labeled in Fig. 1, the table can be supported and leveled.

Preparing for photographs: It is assumed that the teacher who plans to use this system has available a camera and a high-intensity strobe light or, as an alternate, a camera and a rotating disc strobe. Though the results obtained with the rotating disc strobe will not be as good as those taken with the electronic strobe, they will be usable.

The source of illumination should be directional, but with a wide enough spread to cover the area of the table. The axis of the directional beam should be directed to the center of the table and should make an angle of about 20 degrees to the surface. The Dylite spheres should be placed in a salt shaker and sprinkled lightly over the glass.

NOTE: As the spheres reflect light, use only a minimal number. (If the type of puck described here is used, you will find that a very small number of spheres will provide excellent results.)

Making the pucks: Although the pucks can be made of wood, it is recommended that ½" Bakelite be used, because of its greater elasticity. A diagram of a typical puck is shown in Fig. 2. The 1/32" incline cut around the bottom edge of the puck, as illustrated, is especially important. It allows the puck to slide over the spheres without pushing them out of the way. (I have used pucks varying in diameter from 1 inch to 10 inches without problems.)

Fig. 2

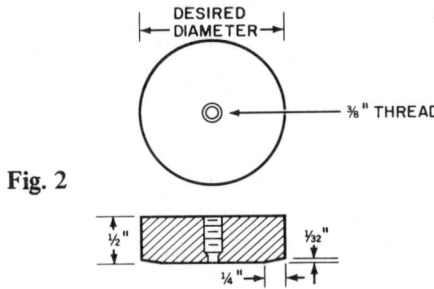

A Student Exercise

An interesting student exercise using this system can be made by constructing a three-mass system with three round pucks of 1", 1½", and 2" in diameter, respectively. The three discs are then connected with light brass brazing rods. Solderless terminals are used for connecting the rods to the pucks (Fig. 3). By allowing the hole diameter of the terminal to be slightly larger than

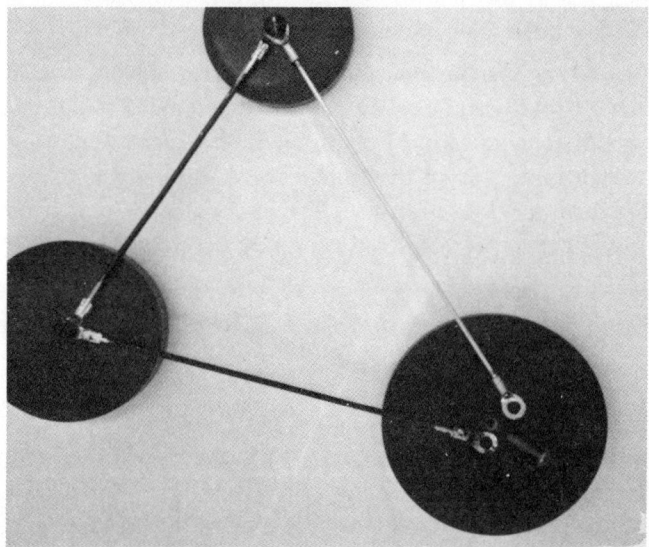

Fig. 3

the bolt in the puck, each of the three pucks has enough freedom of movement to lie flat and still retain rigidity of position.

Procedure: Students can carry out this procedure:

(1) Place the three-mass system on a sheet of graph paper, trace an outline of the three discs, and locate the center of each disc on the graph paper.

(2) Set up an X, Y coordinate system, assuming that the center of the disc represents its center of mass. (The student will now have three sets of X, Y coordinates to locate the centers of masses of the three discs.)

(3) As the mass of each disc will be proportional to its diameter squared, calculate the center of mass of the three-body system by using this equation:

$$\bar{X} = \frac{D_1{}^2 X_1 + D_2{}^2 X_2 + D_3{}^2 X_3}{D_1{}^2 + D_2{}^2 + D_3{}^2} \cdot \cdot$$

(Where D is the diameter of each disc, X the location of the center of the disc on the X coordinate, and $\bar{X}$ the position of the center of mass of the system on the X coordinate.)

(4) Repeat this computation to locate the $\bar{Y}$ for the system, and plot the position of the center of mass of the system on the graph.

(5) Cut out the triangle formed by these centers of the discs and trace this triangle on a sheet of black paper.

(6) Place a white dot at the calculated center of mass, as located on the graph, to locate the center of mass of the system for multiple-flash photography.

NOTE: A collision of the three-mass system and a single puck will result in a photograph similar to that in Fig. 4. It is obvious that the center of mass of the system travels in a straight line after the collision. From this photograph the student can also determine to what degree linear momentum is conserved.

Demonstrating angular collisions: To observe angular collisions, a cut-down "plumber's tool" can be used as the pivot point. The handle should be cut off so that about ½ inch protrudes above the rubber suction cup (Fig. 5). A hole should be drilled in the top of the handle and a nail inserted to act as a pivot. Brass brazing rods can be used to keep the pucks at their proper radius. Fig. 6 shows a typical angular collision of two masses; the crossed puck is incident and the dotted puck the target.

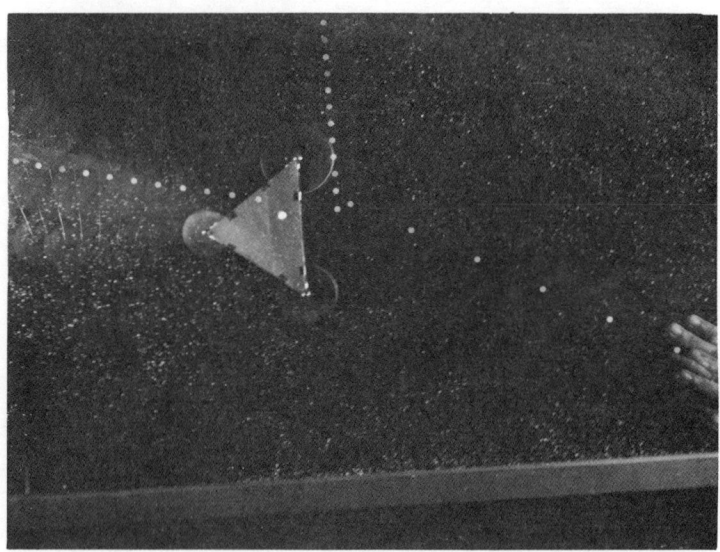

Fig. 4

Fig. 5

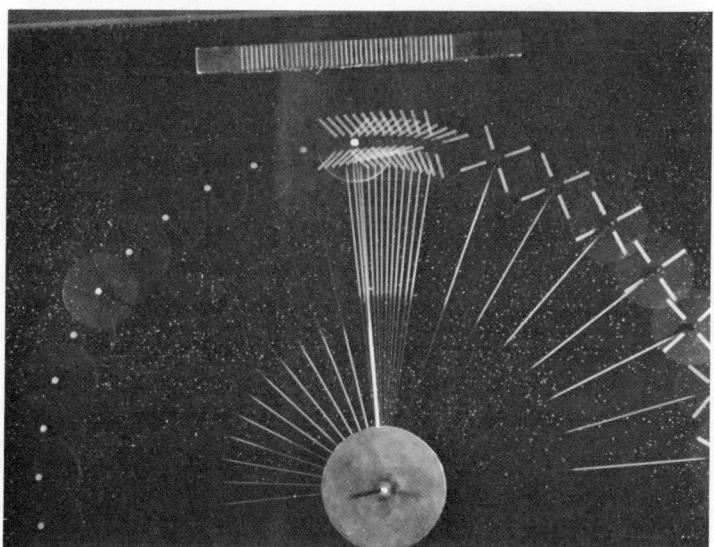

Fig. 6

DERIVATION OF
THE RELATIVITY RATIO

by Duane Hughes

Cyprus High School, Magna, Utah

Experience with the new developmental programs, such as PSSC, shows that student enthusiasm can be developed and maintained with relatively simple equipment and rather fundamental demonstrations and experiments. The task of the teacher, then, becomes that of carefully and effectively applying simple activities to develop concepts which appear, to the student, to be complex and beyond their experience and ability.

One such concept is the frame of reference phenomenon, with attendant ideas of time dilation and mass expansion at velocities approaching the velocity of light. If the textbook does not bother to go beyond the descriptive stage, and most of them do not, the student is inclined to treat the topic as a curiosity. And, of course, sophisticated experimentation in time dilation and mass expansion is impossible in the high school laboratory. The task of the teacher, then, is to develop a demonstrative analogy and derive the formulas involved in the demonstration.

I have found this demonstration to be convenient in achieving these ends.

1. String a white polystyrene ball loosely on a dark string, and adjust it to fall at a uniform velocity. Load a Polaroid camera with 46-L film, which makes a transparency. Use a stroboscope light or uniformly rotating slit stroboscope to get a picture of the falling ball, as shown in Fig. 1

NOTE: Set the camera so that the ball is the obvious feature of the picture, and minimize the background. Have the falling ball appear at the left side of the transparency, and do not use a free-falling ball.

2. Mount the ball-and-string arrangement on a movable cart which can be pulled at a uniform velocity, somewhat less than that of the falling ball. Place a second ball at the bottom of the string. With the camera remaining fixed and the stroboscope operating, pull the cart along. This will result in a transparency similar to Fig. 2.

3. Superimpose the two transparencies and place on an overhead projector. The student will readily see the formation of a right triangle, as shown in Fig. 3.

4. Now suggest to the class that Fig. 1 depicts a beam of light moving from the ceiling to the floor of a rocket ship traveling at near the speed of light, as seen by an observer *on* the rocket ship. The diagonal, or hypotenuse, shown in Fig. 2, depicts the beam of light as seen by an observer who is *outside* the rocket ship watching it go by. Clearly, the diagonal is longer than the vertical path, but since both beams are really the same, and they travel at the same velocity, we can only account for the difference in lengths by recognizing that time is moving more slowly for the observer on the rocket ship than for the fixed observer.

Since the idea of time dilation runs counter to the preconceived notions of most students, a mathematical analysis of the triangle in Fig. 3 can be very convincing. Let S_1 be the vertical displacement of the beam of light; let S_2 be the diagonal displacement of the beam of light; and let S_3 be the horizontal displacement of the rocket ship. Then if C = speed of light, T_1 equals time on the rocket ship, and T_2 equals time for the fixed observer, and V equals the speed of the rocket ship, from the fundamental distance-time-rate formula:

$$S_1 = CT_1$$
$$S_2 = CT_2$$
$$S_3 = VT_2.$$

Now, by the Pythagorean theorem:

$$(CT_2)^2 = (CT_1)^2 + (VT_2)^2.$$

Solving for T_2:

$$(CT_2)^2 - (VT_2)^2 = (CT)^2$$
$$C^2 T_2^2 - V^2 T_2^2 = C^2 T_1^2$$
$$T_2^2 (C^2 - V^2) = C^2 T_1^2$$
$$T_2^2 = \frac{C^2 T_1^2}{C^2 - V^2}.$$

If we divide both members of the equation by C^2/C^2, which, of course, is equal to unity, we have:

$$T_2^2 = \frac{T_1^2}{1 - V^2/C^2}.$$

Taking the square root of both members:

$$T_2 = \frac{T_1}{\sqrt{1 - V^2/C^2}}.$$

The denominator of the right member of this final equation is known as the relativity ratio. It appears in our analysis of time dilation, in an analysis of length dilation, and in an analysis of mass expansion, as velocities approach the speed of light.

> NOTE: The relativity ratio, as developed in the analysis of time dilation, also appears when analyses are made for mass expansion and length contraction. We do not make these analyses in this demonstration.

By substituting various values for V in the explanation, the students can understand the model we have made of relativity, which gives them an idea of the reality of the relative nature of these phenomena.

Fig. 1

Fig. 2

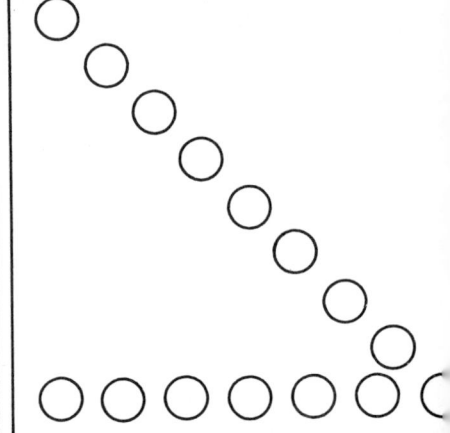

Fig. 3

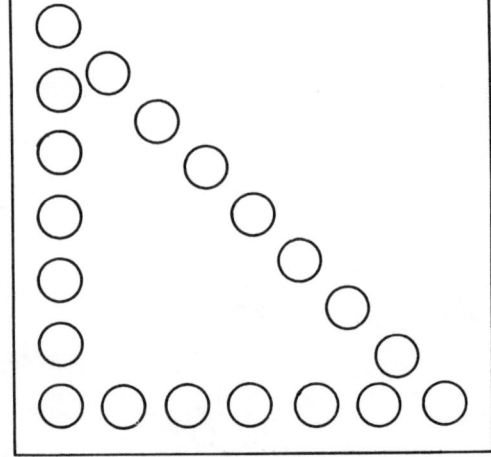

INSTANT PICTURES
IN TEACHING PHYSICS

by Frank W. Gilleland

Maryvale High School, Phoenix, Arizona ·

One of the physics teacher's greatest problems is that the source of data for phenomena he is demonstrating is often transitory. While the "trained eye" of the teacher "sees" the data, the "eye" of the student draws a blank. A solution to this problem is provided by use of the Polaroid or instant camera. The modern Type 47-3000 film makes it possible to get fixed data prints of transitory data; to measure the data at leisure; and to refer to it time and time again.

Making Transitory Data Prints

To take pictures of transitory data, there are two essential accessories in basic work: a *good* camera tripod and either a stroboscope or a timed flasher.

The stroboscope: The Stansi Xenon type of stroboscope is more satisfactory than the neon type because its whiter light makes better print results. By means of a variable-speed disc rotator and a Veedor type of rotation counter, the stroboscope can be calibrated.

NOTE: An excellent class demonstration experiment is to put on low flashing speed and ask students to count the flashes and the time per flash. It is usually found that their results are far from correct. In this way, students discover the fallibility of humans as a routine measuring instrument. The calculation can be repeatedly proven.

Constructing a timed flasher: If a stroboscope is unavailable, you can easily construct a timed flasher (Fig. 1) from an old record player and some "junk." Using a variable speed turntable, multiple time sequences are readily determined.

1. Drill a small hole near the edge of the turntable and to this fasten a binding post in an upright position. From the hole of the binding post, extend a small brush of copper or brass Litz wire. This provides a very good contact brush.

2. Cut a hole in a piece of plywood or other insulating material large

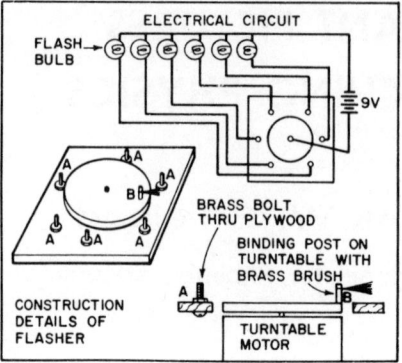

Fig. 1. *Making a timed flasher*

enough to slide over the turntable. Next, drill a series of holes near the edge of the cut-out circle and place brass bolts 8/32" x 1" in the holes so that the brush will strike them as it rotates.

3. Connect a 9-volt transistor radio battery in series with the rotating mechanism and small flashbulbs to the upright terminals; thus, flashbulbs will be flashed as needed and arranged. Raising and lowering the platform will provide an on-off sequence switch for the experiment.

Sequence timing can also be controlled by an on-off switch in series with the battery circuit. For easy visibility, the area of the rotating brush may be painted a bright color. With a little practice, the flashing can be started on any particular contact as the brush rotates. Other accessories will undoubtedly suggest themselves as your use of the camera continues.

Taking Data Prints for the Study of Accelerations and Velocities

In the study of accelerations and velocities, the camera furnishes data in the form of a print which can be compared, measured, and analyzed.

Acceleration of gravity measurements: For example it can be used in determining the acceleration of gravity (Fig. 2).

1. To increase the visibility of the falling object, hang a piece of black felt on the wall in back of the path the object will take.

2. Place the strobe unit or timed flasher in front of and below the pathway of the falling object.

3. Place the camera at a sufficient distance to take in 7 or 8 feet of the pathway of the falling body. The photo is taken on bulb exposure in subdued light, with the strobe light or the flasher as the only source of light.

4. Use a two-position switch (Fig. 3) and an electromagnet to drop a steel ball. By opening the magnetic circuit, the ball is released and the power circuit

Fig. 2

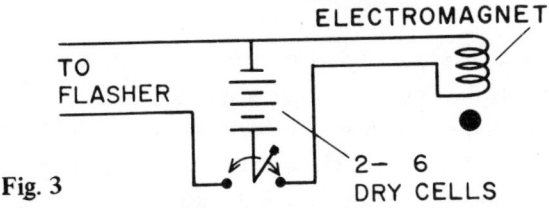

Fig. 3

of the flasher is closed. The amount of error induced in the lack of absolute synchronization during the first free-fall period is so small as to be negligible in the calculations.

5. For the background measuring media, tape together meter sticks. Use vernier calipers or dividers for direct measurements from the pictures. Since the falling object is within 1 or 2 centimeters of the background scale, parallax is not a specific problem.

Errors caused by parallax: An excellent demonstration of errors caused by parallax is made by bringing the falling object ½ meter in front of the scale toward the camera and then comparing readings with a properly placed fall. From a print produced under these conditions, the following data can be obtained if the time per flash is known:

a. Total distance fallen in a given period of time, $S = 1/2gt^2$.

b. Distance fallen in any particular time unit, $s = 1/g(2t - 1)$.

c. Rate of increase (v) in distance traveled per time unit.

d. Effect of Vo or Vi on any time unit.

NOTE: The transition process of changing units of measurement from the photoprint to the actual length of fall as shown on the

print is a valuable technique for students to learn for future indirect measurements.

Uniformly accelerated motion: A control investigation of uniformly accelerated motion can be made.

Nail two 1-inch by 2-inch boards edge to edge as shown in Fig. 4. To

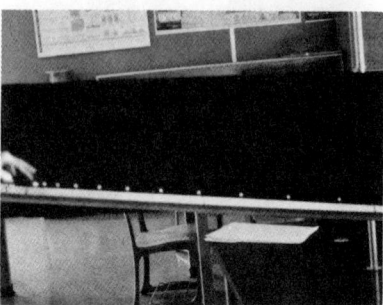

Fig. 4

increase or decrease the rate of acceleration, the inclined plane may be raised or lowered. With a strobe light or flasher suspended in an overhead position, the rate of descent of the rolling ball or other objects can readily and accurately be determined. By varying the angle of the "V" trough, basic trigonometric functions can be made to have practical meaning to students as the rolling component of the gravitational force is determined.

> **NOTE: Varying the size and shape of the rolling objects lends interest to the experimental procedure. Students often use small automobiles in their individual investigations, even to the point of comparing respective coefficients of frictions.**

Prints for Studying Other Phenomena

Wave motion: Use of the Polaroid camera in the study of wave motion is obvious. Permanent records of different wave patterns can be isolated and carefully studied. A small amount of blue fountain brush ink (water soluble type) or blue food coloring added to the water of the ripple tank greatly increases the contrast of the Polaroid wave pictures. All interference and flow patterns are especially adaptable to this picture technique.

Other uses: With a Van de Graaff generator or a Tesla coil, pictures may be obtained that will provide essential data for the measurement of potential and static forces involved in the QQ^1 over r^2 relationship (Fig. 5). In a time exposure, the ionization patterns of the air through which a spark is jumping can be studied. The traced designs of the oscilloscope can also be permanently recorded for detailed study. For this the use of a closeup auxiliary portrait lens and short hood are recommended.

Students may be asked to take pictures of construction projects to seek out applications of the principles of physics in the building industry. Since it is required that a student get the permission of management to take the pictures, many new contacts are opened to him. This has not only helped to stimulate classwork but has also led to better public relations between school and industry and between the school and community as a whole.

Fig. 5

14

USING ELECTRONICS IN THE LABORATORY

DEMONSTRATING
THE PHOTOELECTRIC EFFECT
USING VTVM

by Buford L. Williams

Kimball County High School, Kimball, Nebraska

The high cost (up to $1,000) of commercially available apparatus for firsthand study of the photoelectric effect prohibits its use in most high school physics classes. As described here, however, a suitable demonstration of this phenomenon can be performed with a vacuum tube voltmeter, such as the Heathkit IM-11, and a monochromatic light source such as the Macalester No. 3400.

> ACCURACY: Since the VTVM is not sensitive enough for accurate quantitative work when used in this manner and no account is taken of the counter-current of emission from the collecting wire of the phototube, a high per cent of error can be expected. However, the result of calculating Planck's constant should be of correct order of magnitude.

Background Material

The action of electromagnetic radiation in causing the emission of electrons is called the photoelectric effect. Light of suitable frequency incident upon certain metallic surfaces is responsible for this emission of electrons—called photoelectrons—from these materials.

Heinrich Hertz discovered the photoelectric effect in 1887, during his research on electric waves. He noticed that the sparks produced in the gap of the secondary circuit were affected by light falling upon the gap from the spark in the primary circuit. Subsequently, Hertz found that the ultraviolet light from the primary spark incident upon the negative terminal of the secondary gap was creating this phenomenon.

Later experimenters showed that the effect consisted of the emission of electrons and formulated two empirical laws based upon their observations:

(a)　The first law states that the rate of emission of photoelectrons is directly proportional to the intensity of the incident light, and that the kinetic energy of the photoelectrons is independent of the intensity of the incident light.

(b) The second law states that within the region of effective frequencies, the maximum kinetic energy of the photoelectrons is directly proportional to the frequency of the incident light.

In 1905, Albert Einstein showed how Planck's new quantum theory of radiation could be used to account for the photoelectric effect.[1] The quantum theory states that if all of the energy of a quantum (photon) is given to a single electron, it would be emitted from the surface provided the energy were sufficient to take the electron through the surface. If the minimum energy for escape (called the work function of the material) is W_o, γ is the frequency of the radiation, e is the electron charge (-1.6019×10^{-19} coulomb), and V_{co} is the retarding potential required to stop the fastest photoelectrons, then

$$\tfrac{1}{2}mv^2 = h\gamma - W_o = eV_{co}. \quad (1)$$

Robert A. Millikan[2] had previously shown that a plot of V_{co} as a function of light frequency was linear and that the required potential to stop the photoelectrons becomes zero at some frequency γ_{co}. This cutoff frequency depends upon the particular photosensitive surface and corresponds to the lowest frequency for light which can free electrons from the surface. From equation (1):

$$\text{slope} = \frac{\Delta V_{co}}{\Delta \gamma} = \frac{h}{e}. \quad (2)$$

Description of the Apparatus

Its ability to amplify small currents makes the VTVM an extremely useful instrument for measuring photoelectric currents. It employs vacuum tubes for amplification of measurements to ensure sensitivity and stability. The potentials from the phototube that are applied to the VTVM probe are conducted to one grid of a twin triode amplifier tube. This sets up an imbalance between the two cathode currents of the tube. The meter movement is connected between the two cathodes, the pointer responding proportionally to this current difference.

NOTE: The very small photocurrents in this demonstration allow us to connect the VTVM as an ammeter rather than a voltmeter; the meter movement will correspond to the current through the VTVM rather than the voltage across it.

---

[1] For this explanation Einstein was awarded the Nobel prize in 1921.

[2] Robert A. Millikan (1868-1953) received the Nobel prize in physics in 1923 for verifying Einstein's photoelectric equation and determining the charge of the electron.

Photocurrents from the 929 phototube (Fig. 1) pass through the VTVM. The retarding potential from the 1.5-volt battery through the 75 K ohm potentiometer is applied to the 929 tube in such a manner as to cause an electric field inside the tube that will oppose the motion of the photoelectrons, thus slowing or stopping them. The V_{co} across the tube is measured by the voltmeter (0-1.5 volt) when the photocurrents are just stopped; i.e., when the VTVM reads zero current.

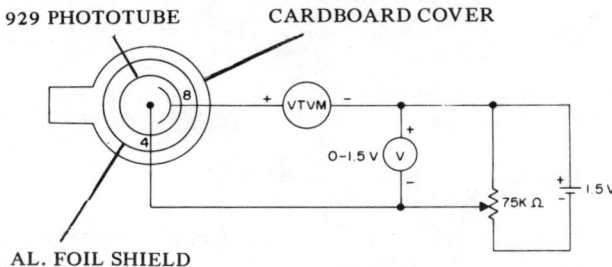

Fig. 1: Phototube Circuit

IMPORTANT: The 929 phototube should be shielded from the external static electricity by covering it with aluminum foil having a hole to admit light to the photosensitive surface. This foil should, in turn, be covered with cardboard to exclude all light except from the light source.

Demonstration 1

Demonstration of the first law can best be done in a semi-darkened room using about six standard candles set approximately 1 foot from the phototube as a light source. As the intensity of the source is changed by lighting successive numbers of candles, the maximum current is read from the VTVM and the V_{co} is determined by adjusting the potentiometer so that the VTVM reads zero current (Fig. 2).

A straight-line graph of intensity vs current (Fig. 3) shows that the current is directly proportional to the intensity of the source. The V_{co} will remain essentially constant for each of the source intensities. This indicates that the fastest photoelectrons in each trial all possess the same energy, for it takes the same retarding potential to stop them.

Demonstration 2

Monochromatic light is necessary to demonstrate the second law. The

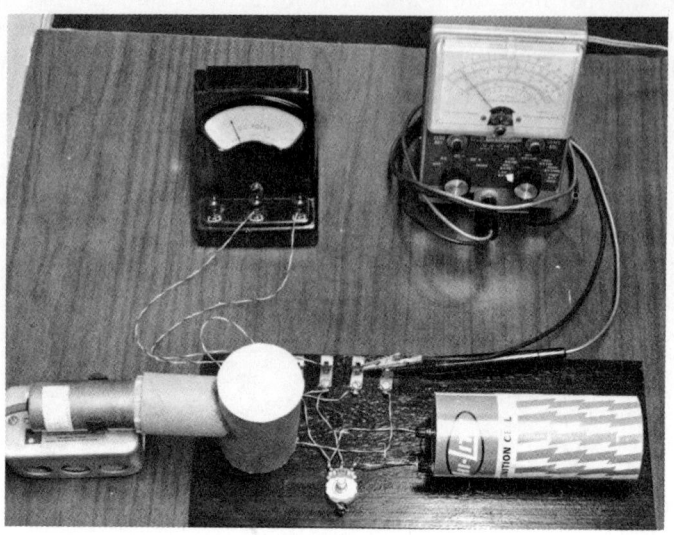

Fig. 2

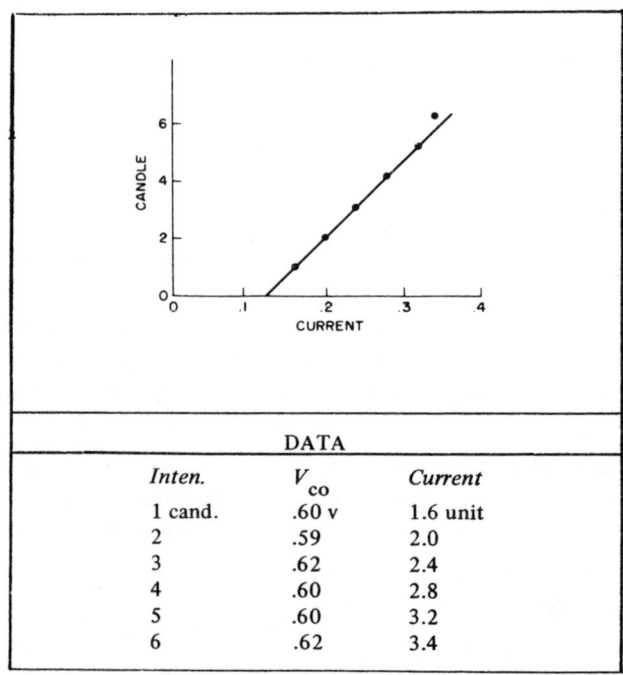

DATA

Inten.	V_{co}	Current
1 cand.	.60 v	1.6 unit
2	.59	2.0
3	.62	2.4
4	.60	2.8
5	.60	3.2
6	.62	3.4

Fig. 3

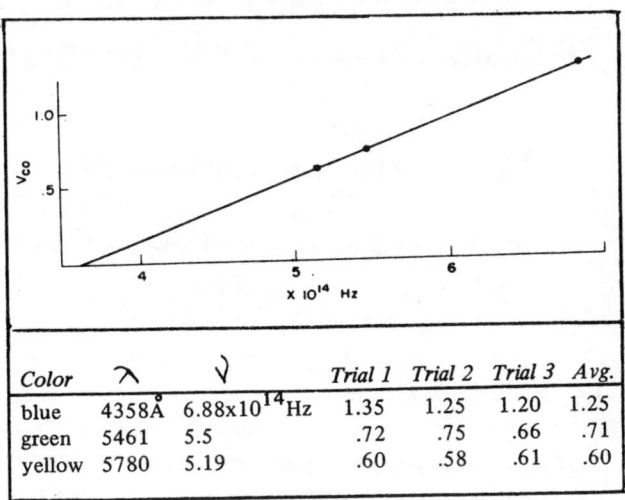

Color	λ	ν	Trial 1	Trial 2	Trial 3	Avg.
blue	4358Å	6.88×10^{14} Hz	1.35	1.25	1.20	1.25
green	5461	5.5	.72	.75	.66	.71
yellow	5780	5.19	.60	.58	.61	.60

Fig. 4

Macalester source is supplied with three sets of filters to isolate the three bright lines in the visible spectrum of the mercury discharge bulb; a green filter for the 5461 Å line, a pair of orange-red filters to isolate the 5770-5790 Å yellow lines, and a blue filter to isolate the 4358 Å blue line (Fig. 4).

WARNING: *THE MERCURY LAMP PRODUCES STRONG RAD-IATION IN THE ULTRAVIOLET. CAUTION YOUR STUDENTS NOT TO LOOK AT THE BARE BULB.*

To demonstrate the second law, the V_{co} is found for the three monochromatic light frequencies. For the best possible plot of V_{co} vs ν, it is well to take an average of several trials for V_{co}. Using the data for this demonstration and equation 3, it is possible to get an approximate value of Planck's constant, h,

$$h = \frac{\Delta V_{co}}{\Delta \nu} \times e \qquad (3)$$

$$h = \frac{(1.25 - .60) \text{ joule/coul}}{(6.88 - 5.19) \times 10^{14} \text{ Hz}} \times 1.6 \times 10^{-19} \text{ coul}$$

$h = 6.2 \times 10^{-34}$ joule-sec (6.62×10^{-34} joule-sec accepted value).

A VARIATION ON THE
THOMSON e/m EXPERIMENT

by David D. Lockhart

Allen Park High School, Allen Park, Michigan

One of the classic demonstration experiments in modern physics is the J.J. Thomson determination of the charge to mass (e/m) ratio of the electron. Though the apparatus necessary to stage this demonstration can easily be obtained from a scientific supply house, in most cases its cost is considerable. However, the experiment can be successfully duplicated using an oscilloscope, which is generally already available in the physics laboratory.

Apparatus and Procedure

To execute the experiment with the oscilloscope, a coil to furnish the magnetic field must be built. The coil form should be made from wood lath (about 1" x 1/4") and its inside dimensions cut so that the form will fit over the scope snugly and extend about 1 foot in front of and behind the instrument (Fig. 1). For added strength, the corners of the form can be reinforced with rectangular blocks of wood. A coil of ten turns of #20 enameled wire should then be wound around the outside of the prepared form.

(1) As the first step in setting up the experiment, remove the oscilloscope from its housing. This can usually be accomplished by removing two screws at the 'rear of the instrument. The scope can then be withdrawn from its case, being careful not to damage the glass cathode ray tube (CRT).

(2) Once the case has been removed, place the coil over the scope and support it at a height on a level with the axis of the CRT. Connect a 6-volt battery to the coil through a 20-ohm potentiometer so the current flow through the coil can be regulated. Monitor this current flow with a 0 - 5 ampere DC ammeter (Fig. 2).

(3) With the power to the oscilloscope disconnected, insert a DC voltmeter between the cathode and the accelerating anode of the CRT. This voltage is customarily about 1200 volts.

WARNING: DO NOT TOUCH THE INTERNAL PORTIONS OF

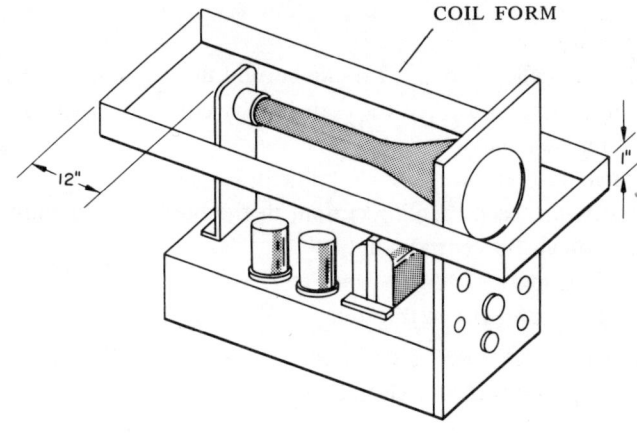

COIL FORM

12"

1"

Fig.1

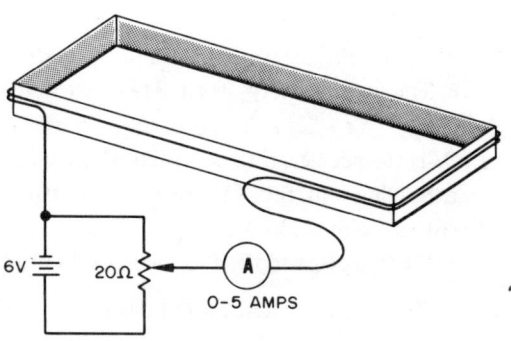

6V   20Ω   A   0-5 AMPS

Fig. 2

**THE SCOPE WHEN THE POWER IS ON. THIS VOLTAGE IS
DANGEROUS.**

(4) With the scope on and the vertical and horizontal drives turned off or
at a minimum, establish a well-defined spot on the center of the screen using the
focus, contrast, and horizontal and vertical position controls.

(5) As the coil is energized, the spot on the screen will jump upward or
downward depending upon polarity of the field. Record the coil current (I) and
measure the deflection of the spot (X).

Calculations

The calculation of the e/m ratio proceeds as follows:

a. The magnetic induction established at a distance from a single straight wire is:

(1) $B = \mu_0 I/2\,R,$

where B is the induction in webers/meter^2; I is the current in amperes; R is the distance from the wire in meters; and μ_0 is the permeability constant for air and is 12.57×10^{-7} webers/ampere-meter.

b. Since the magnetic field in this experiment is generated midway between two bundles of wires, each carrying current in the opposite direction, the total magnetic induction at this central point is:

(2) $B = 2(\mu_0 IN/2\pi R),$

where N is the number of turns of wire in the coil.

c. When the electron beam is deflected by the magnetic field, the radius of the arc through which the beam is bent is given by:

(3) $r = (S^2 + X^2)/2X,$

where S is the distance from the anode of the CRT to the screen in meters and X is the deflection of the spot on the screen in meters.

d. The charge-mass ratio of the electron is:

(4) $e/m = v/Br,$

where v is the velocity of the beam in meters/second.

e. The kinetic energy formula can be written:

(5) $Ve = \frac{1}{2}mv^2,$

where V is the accelerating potential of the electron beam in volts.

f. Solving equation (5) for v and substituting in equation (4), the result is:

(6) $e/m = 2V/B^2 r^2.$

g. Since V, B, and r are known, a value for e/m can be calculated.

TABLE 1: Table 1 shows typical results.

Trial	I (amps)	X (m)	r (m)	B ($\times 10^{-5}$ w/m^2)	e/m ($\times 10^{11}$ coul/kg)
1	1.00	.0120	3.50	3.28	1.73
2	1.25	.0160	2.63	4.10	1.98
3	1.50	.0185	2.28	4.92	1.83
4	2.00	.0250	1.69	6.56	1.87
5	2.30	.0290	1.46	7.55	1.89
6	2.50	.0315	1.35	8.20	1.88
7	2.60	.0320	1.33	8.54	1.78

In carrying out this experiment, the following values were established:

$S = .290$ meters
$R = .122$ meters
$N = 10$ turns
$V = 1150$ volts

The calculations for the first set of trial data from Table 1 produces:

(2) $\quad B = \dfrac{2\mu_0 IN}{2\pi R} = \dfrac{2(12.57 \times 10^{-7})(1.00)(10)}{2(3.14)(.122)} =$

$$3.28 \times 10^{-5} \text{ w/m}^2$$

(3) $\quad r = \dfrac{S^2 + X^2}{2X} = \dfrac{(.290)^2 + (.012)^2}{2(.012)} = 3.50 \text{ m}$

(6) $\quad e/m = \dfrac{2V}{B^2 r^2} = \dfrac{2(1150)}{(3.28 \times 10^{-5})^2 (3.50)^2} =$

$$1.73 \times 10^{11} \text{ coul/kg.}$$

The accepted value for e/m is 1.76×10^{11} coulombs/kilogram.

NOTE: While this approach does not provide an extremely accurate method of determining the charge-mass ratio of the electron, the procedure is straightforward and illustrates several important scientific principles.

SOUND TRANSMITTED ON A LIGHT BEAM

by Kenneth A. Wright

Central Michigan University, Mount Pleasant, Michigan

There are many practical uses in everyday life for photoelectric cells. In industry, they are used for counting, sorting, regulating artificial light, protecting workers, and performing many other tasks. In the home, they are applied to opening doors, turning on lights, announcing guests, and activating fire and burglar alarms.

NOTE: Following are directions for an interesting and easily staged

demonstration of sound transmitted on a light beam. This demonstration effectively illustrates a great deal of science in a very simple way.

Background material: The photoelectric effect was discovered in 1869 by Hallwachs, who found that crystals of fluorite became electrically charged not only by heat, but also by exposure to sunlight or to light as an electric arc. Both sunlight and the carbon arc are rich in ultraviolet light. All photoelectric devices convert light or radiant energy into electrical energy, employing various methods of conversion.

Photoelectric cells are classified as follows:

> *Photoemission:* In photoemission cells, the energy of the light beam causes emission of electrons from a metal surface enclosed in an evacuated or gas-filled bulb.
>
> *Photovoltaic:* In photovoltaic cells, the radiant energy causes generation of an emf. The resulting current may be made proportional to the radiant energy. These cells consist of oxides or semi-conducting layers on metal base plates and need not be evacuated.
>
> *Photoconductive:* In photoconductive cells, the electrical resistance varies in accordance with the radiant energy received. Originally made of selenium deposited on a glass plate, these cells now employ a semiconductor junction.

Demonstration Setup: The Sound-on-Light Beam System

Various setups can be used to illustrate light beam communication. The following system not only works well but has the special advantage of requiring a minimum of simple equipment.

Apparatus and assembly: To conduct this demonstration, you will need:

2 amplifiers
1 flashlight with batteries
1 photoelectric cell

These materials comprise the three parts of the sound-on-light system: (1) a light source which can be intensity-modulated by sound waves—the transmitter; (2) a pickup unit to convert the modulated light back into sound—the receiver, and (3) an optical system to carry the light from the transmitter to the receiver—the photoelectric cell.

1. Transmitter: A low-power amplifier of about 3 to 5 watts will suffice for the transmitter. The one we use is a two-stage amplifier that was built in the laboratory, but a small-gain amplifier with a record player or microphone attached to the input should be effective.

NOTE: Most radio receivers should have sufficient output to be employed for this purpose

To set up the transmitter, attach the flashlight rather than a speaker to the output, as in Fig. 1; that is, connect the secondary of the output transformer in series with the flashlight batteries. (As the signal comes from the output, the light will increase and decrease with the strength of the signal. The light is very sensitive and will respond to high frequencies.)

CAUTION: The volume control on the amplifier must be carefully adjusted or the flashlight bulb will be burned out.

2. Receiver: The receiver (Fig. 2) requires an amplifier with more gain than the transmitter. A three-stage amplifier or perhaps a radio that has provision for a record player or a microphone can be used.

3. Photoelectric cell: The photoelectric cell should be connected to the receiver in the same way and same place as the record player or microphone would be connected, with one end to the grid of the tube and the other end grounded (Fig. 2).

We have had good results using a selenium photovoltaic cell in this demonstration, though a silicon photovoltaic cell would be even more successful. Both types of cells are inexpensive and can generally be purchased for about a dollar.

PROCEDURES: To demonstrate the principle of sound transmitted on a light beam, the light from the flashlight is focused on the photoelectric cell. Modulating the flashlight illustrates the same principle as modulating a radio wave on an AM radio. It also illustrates the principle of the photocell as more light falling on the cell produces more voltage, which in turn produces more signal in the amplifier.

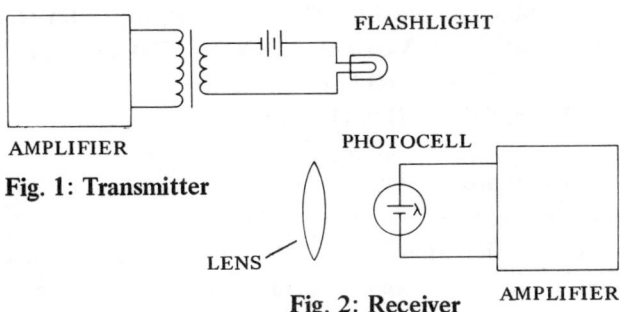

FLASHLIGHT

AMPLIFIER

Fig. 1: Transmitter

PHOTOCELL

LENS

Fig. 2: Receiver   AMPLIFIER

FREQUENCY MEASUREMENT
WITH A SCALER

by C. Kenneth Baker

Longmont Senior High School, Longmont, Colorado

Most physics teachers are probably familiar with the scaler in conjunction with the Geiger counter. The scaler offers better statistical estimates of radiation than a meter read-out, due to the longer time base. But this apparatus has many useful applications besides that of counting radiation. In fact, the opportunities for its effective use in the physics class are almost limitless, and the teacher will discover that the electronic scaler's flashing lights are almost hypnotic in holding students' attention.

Obtaining a scaler: Many scalers are now available commercially, either as integral parts of Gieger counters or as separate instruments. When shopping, don't overlook the surplus market, which often provides some excellent buys. (The scaler I'm using cost $2.00.) While it is true that the surplus scaler will most likely be a bulky vacuum tube model with perhaps a binary readout, this is no great disadvantage. The binary scale may even be desirable for demonstrating a base 2 counting scheme, a "computer" demonstration which particularly appeals to junior high school students.

Using a Geiger Counter Scaler

WARNING: IF YOU ATTEMPT TO USE A SCALER WHICH IS AN INTEGRAL PART OF A GEIGER COUNTER, BE CERTAIN TO ELIMINATE THE HIGH VOLTAGE (500 to 2500 VOLTS) USED FOR THE G-M TUBE. TO ACCOMPLISH THIS, PULL THE HIGH-VOLTAGE RECTIFIER TUBE.

Applying the signal: After elimination of the high-voltage danger by removing the high-voltage rectifier tube, a signal can be applied to the counter through the G-M tube plug. However, the counter is still unlikely to count unless the pulses sent in are very brief, such as those from the strobe light and photocell. Ordinary sine waves are not likely to trigger such a circuit. What is needed is a wave-shaper to convert the waves into spikes or pulses with a short rise time.

Converting the pulse: The circuit shown in Fig. 1 provides an economical way of converting any pulse of suitable amplitude into a countable pulse. Of course, if you are fortunate enough to get a scaler designed for general counting or frequency measurement, the modifications are unnecessary.

This circuit is sensitive yet stable and is capable of converting a sine wave of 1 volt peak to peak into a square wave output of about 10 volts peak to peak. The rise time is less than 1 microsecond, and it possesses a usable output in excess of 200 kilo-Hertz. For low frequency sine wave inputs (less than about 500 cps), it may be necessary to increase C_1 to 0.5 mf to provide sufficient drive.

The component values are not very critical, but R_1 and R_2 should be matched. Q_1, Q_2, and Q_3 are pnp type 2N404, but a number of substitutes will work as well. R_3 is variable to provide an exact balance between the conduction states of the flip-flop and is not a gain control. The adjustment is somewhat critical for some inputs. Diode D_1 may prove helpful if the counter counts at double the input frequency.

Using a selenium photocell: To provide adequate sensitivity from a selenium photocell input, an amplifier stage is shown in Fig. 2 and Photo 1. A common battery supply is suitable. Surprisingly, it is difficult to trigger the

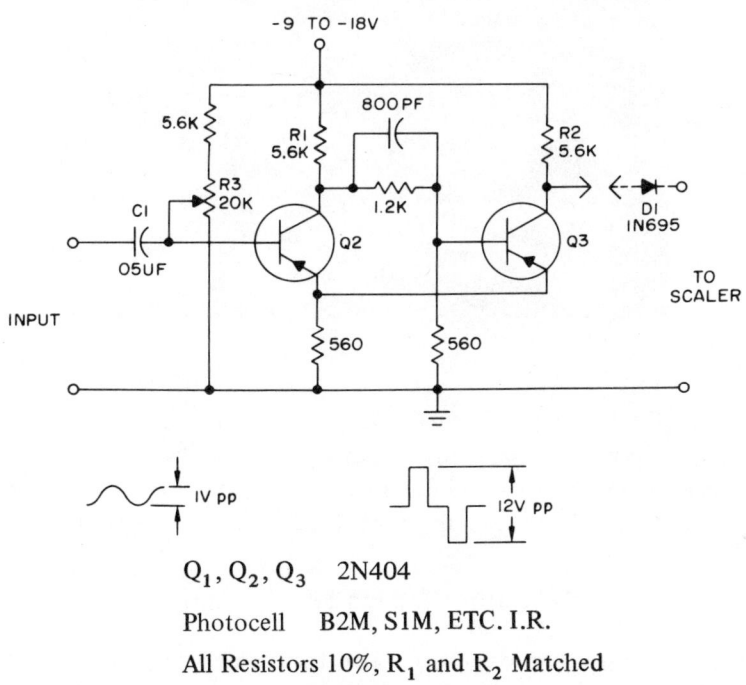

Q_1, Q_2, Q_3 2N404

Photocell   B2M, S1M, ETC. I.R.

All Resistors 10%, R_1 and R_2 Matched

Fig. 1: Transistorized Schmitt Trigger

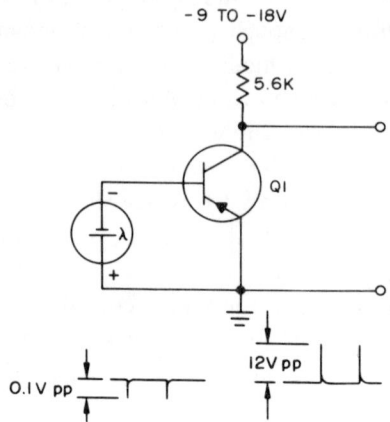

Fig. 2: Photocell Amplifier

Photo 1

scaler manually because a manual switch transmits a number of pulses on opening and closing. Consequently, using a telephone dial for an input switch usually leads to an unreliable count. One way to overcome this problem is to use the switch to trigger the strobe tube manually and use the photocell input.

Measuring Camera Shutter Time

One demonstration experiment with the scaler, which can be performed early in the year for your physics class, involves measuring the time a camera shutter is open and comparing the results with the shutter markings. Students find this experiment interesting in itself, though its real purpose is to introduce them to the nature of measurement, significant digits, and the use of "black boxes."

NOTE: The demonstration has a purpose similar to that of the PSSC film "Measurement," and it possesses much the same flavor.

Procedure: To conduct the experiment:

(1) Place the shutter of the camera on "time" and adjust a strobe light to flash about 100 times per second through the lens.

(2) On the opposite side of the lens, set up a photoelectric cell to pick up the flashes and convert them to electrical pulses, which are then counted on the scaler (Fig. 3 and Photo 2).

(3) After the function of each "black box" is understood by the students, take a count of the strobe flashes for perhaps ten seconds to determine the flash rate by actual count.

NOTE: Let the students decide the time interval to use. Their intuition will tell them that it cannot be too short or too long, and it is part of the experiment's purpose to develop this sense.

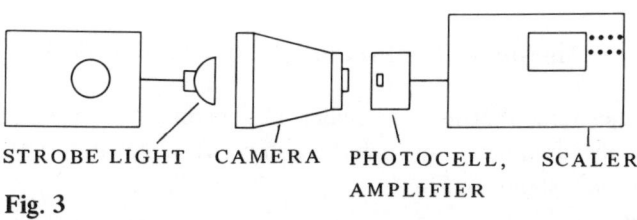

STROBE LIGHT   CAMERA   PHOTOCELL,   SCALER

AMPLIFIER

Fig. 3

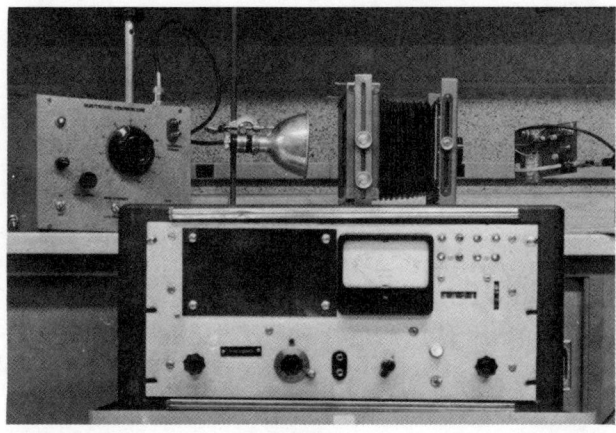

Photo 2

For simplicity, the strobe may be adjusted to as near 100 flashes per second·as possible. It will become evident that exact repeatability is not possible; reasons for this will need to be explored, and some system of notation will have to be developed to reflect this fact, for example, 100 ± 1 cps.

(4) With the flash rate established with a known tolerance, the shutter can be set to open for one second, one-fifth second, etc., and the accuracy of the shutter can be calculated.

Analyzing results: It is not uncommon to find well-made shutters with 50% errors, but this is not really the point of the experiment. As the shutter speed is increased to 1/100 second, it becomes evident that a higher flash rate is desirable. The number of bits of information (discrete flashes) is too small to make a useful measurement; the number of significant digits has dwindled to nearly nothing. The time interval, $\triangle t = (t_2 - t_1)$, gives $t = [(101 \pm 1) - (100 \pm 1)]$, which is clearly an uncertain result.

Follow-up: This experimental data provides a good intuitive introduction to information theory, with the concept that a measurement involves a series of yes-no decisions. The more decisions which can be made of this kind concerning a measurement, the more information one has and the more significant digits he is entitled to use.[1]

Measuring the Frequency of a Tuning Fork

Another somewhat more complicated experiment with the scaler involves measuring the frequency of a tuning fork. The apparatus setup for this demonstration is shown in Fig. 4.

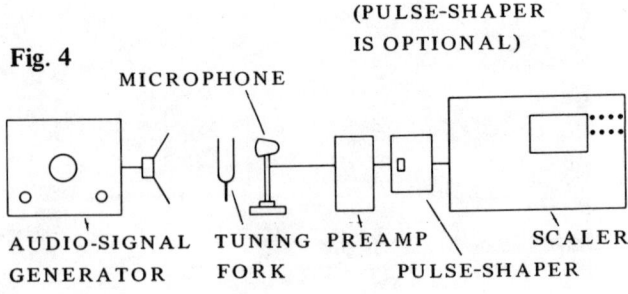

Fig. 4

(PULSE-SHAPER IS OPTIONAL)

MICROPHONE

AUDIO-SIGNAL GENERATOR   TUNING FORK   PREAMP PULSE-SHAPER   SCALER

---

[1] See Physical Science Study Committee, *Physics.* (Boston: D.C. Heath & Co., 1965) Chapter 10.

Procedure:

(1) Place the tuning fork in front of a microphone and amplify its output through a preamplifier (a good Hi-Fi amplifier will work).

(2) Feed the signal to the scaler (Fig. 4). (Acutally, the signal is fed first to the wave-shaper and then to the scaler as previously explained.)

Since the tuning fork has a limited period of useful vibration, it may be difficult to get many significant digits into the frequency measurement. In this case, synchronize the tuning fork with the earlier calibrated audio-signal generator connected to a speaker. Then either read the frequency from the dial on the signal generator, or again count the frequency from the speaker. The latter should be more accurate.

If you have difficulty hearing beats between the two signals, you may wish to use an oscilloscope to form a one-to-one Lissajous pattern. The oscilloscope can also be employed to extend the usable frequency response of the system by using Lissajous patterns to provide other than one-to-one ratios; for example, 10:1.

OTHER EXPERIMENTS: Many further experiments using the scaler are possible. It has been mentioned, for instance, that the stroboscope can be directly calibrated with the scaler. The same is true for other electronic gear, such as an audiosignal generator. In each case, the square wave output is fed into the scaler for a suitable time interval and the frequency calculated.

Index

Work and energy (*cont'd.*)

> data and graphs plotted, typical, 199-201
> displacement, force parallel to as work and energy, 203-205
> mechanical energy, analysis of, 201-203
> thermodynamics, introducing, 192-195
> > entropy, concept of, 193-194

Work and energy (*cont'd.*)

> first law of physics and first law of nature, 194-195
> thermal capacity of water, 192-193

Z

Zacharias, Dr., 188
Zemansky, Mark W., 63, 104n, 136, 240